INTRODUCTORY ALGEBRA:

A College Approach

By the same authors

ARITHMETIC: A COLLEGE APPROACH (*1966*)

INTRODUCTORY ALGEBRA:

A College Approach

MILTON D. EULENBERG
THEODORE S. SUNKO

Wilbur Wright College
Chicago, Illinois

SECOND EDITION

JOHN WILEY & SONS, INC., *New York · London · Sydney*

Preface to the Second Edition

The second edition of *Introductory Algebra* retains the basic features which have contributed to the success of the first edition. Indeed, the reaction of many mathematics teachers to the overemphasis of "modern mathematics," with its extensive use of set notation and symbolism, has convinced the authors and editors that only a moderate revision is called for. We believe, as we stated in the preface of the first edition, that the degree of rigor must be commensurate with the ability and experience of the students for whom this book is intended, and that unnecessary complications in the text as well as in the problems should be avoided.

The major changes in the text consist of the addition of two sections on inequalities and a more complete development of the number system. New sections on rational and real numbers have been introduced and a postulational approach to number fields has been given greater emphasis. A number of minor changes have been made in the text as well as in the problems wherever it was thought that a better understanding of the material under discussion would result. The chapter on sets has been left in its original place as a final chapter because we believe, as do many of the users of the book, that, as we stated in the Preface of the First Edition, " . . . the student will be in a better position to understand the operations and algebra of sets after he has mastered the fundamentals of algebra as we have developed them."

To the users of the first edition, who have suggested many of the changes we have made, we are grateful. We are indebted to the publisher and its editorial staff for giving us the opportunity to serve the college student, for whom this book is still written.

Chicago, Illinois
April, 1967

Milton D. Eulenberg
Theodore S. Sunko

Preface to the First Edition

Many college students have inadequate or insufficient preparation in algebra for college-level work in mathematics. *Introductory Algebra* is intended for students in this group who can attain in one semester the preparation necessary to enter college algebra or an equivalent-level mathematics course. Since the book is designed for college students, it assumes a degree of maturity and sophistication on the part of the user somewhat above that characteristic of the usual student of elementary algebra. This assumption is reflected in the organization, method, and scope of the book.

We believe that college students can be exposed to longer sequences of related ideas than most high school students. Consequently, the expository material and exercises are organized somewhat on the unit plan with each section designed to cover approximately one lecture period. However, the unity of a particular section is not sacrificed in favor of adhering to a rigid time schedule and, therefore, the time for each may vary as the instructor desires.

We believe also that the concepts and techniques of algebra gain in meaning if they are based on and developed from fundamental principles. Therefore an effort is made to present algebra as a logical structure, founded on undefined terms, defined terms, and postulates, and developed by reasoning and proof from certain assumptions. The postulates governing the fundamental operations are not only stated, but they are also incorporated in the exercises as a means of making the student think about them, work with them, and understand them.

Throughout the book we have tried to be as precise and rigorous as possible, but we have also kept in mind the level of ability of the student for whom this book is intended.   The order in which the topics are introduced is designed to permit early application of algebraic techniques. The exercises have been carefully constructed to develop proficiency in working with algebraic expressions and to promote understanding of the principles involved.   Unnecessarily complicated problems have been avoided.   Special attention has been devoted to the construction of problems which may lead the student to investigate relations and encourage him to raise questions or initiate discussion.

The discussion of functional relations is given considerable emphasis. A particular effort is made to relate algebraic and geometric systems of representation and to deal with meaningful elementary applications of the function concept.   We recognize, of course, that such attempts are necessarily restricted by the degree of abstractness permissible or prudent in a book of this type.

The chapter on variation serves as an excellent summary of some of the principles of algebra and functional relations.   It has the further merit of being readily applicable to other fields of study, particularly the sciences.   Since such applications frequently involve measurement and computation, the chapter on variation is preceded by chapters on approximate numbers and logarithms.

The concept of set appears early in the book as a means of introducing the natural numbers and their properties.   It is referred to in subsequent sections—particularly in the extension of the number system, the discussion of a variable, the definition of a function, and the solution of certain equations.   However, we feel that the student for whom this book is intended would find too abstract such concepts as, for example, the use of cartesian sets as a foundation for the study of simultaneous equations or functional relations.   We have therefore deferred the more formal work with sets until the last chapter, where the teacher may use it or omit it as he wishes.   We believe that the student will be in a better position to understand the operations and algebra of sets after he has mastered the fundamentals of algebra as we have developed them. In turn, those students who can handle the material of the final chapter will find a fresh point of view, an increased appreciation of the nature of algebra, and a deeper understanding of the concepts developed in this book.

Milton D. Eulenberg
Theodore S. Sunko

Chicago, Illinois
November, 1960

Contents

INTRODUCTORY ALGEBRA:
A College Approach

CHAPTER 1

Introduction—the Meaning of Algebra

What Is Algebra?

The nature of algebra has been described in various ways. Algebra has been called the cornerstone of modern mathematics, a statement which declares its importance but tells us nothing about it. Algebra has been described as an extension of arithmetic, similar to arithmetic because it deals with numbers and different from arithmetic because it stresses generalizations. Algebra has been called a language because its symbols take the place of words. Algebra has been vaguely defined as a method of mathematical computation in which letters and other symbols are employed. None of these statements, nor perhaps any statement, can describe exactly what algebra is, any more than a single statement can describe exactly what a human being is. This book will treat algebra as a language, as a method of problem solving, and as a logical structure; an understanding of its meaning and its methods will gradually develop as the study of the book progresses.

Algebra as a Language

Mathematics, as a language for communicating certain kinds of ideas, is probably the clearest language invented by man. This is largely because the symbols used in mathematics are admissible of more precise definition than even the common words of our everyday language.

Consider, for example, the difficulty of defining so common a word symbol as "book." We could begin by including under this concept any object which is a collection of sheets, bound along one edge, and protected by some covering material. In this sense it includes not only a novel but also a checkbook, album, or notebook. Further reflection might reveal

1

that this concept of "book" would exclude certain types of records which by virtue of their function should certainly be called books.   For example, the "rolls" of ancient Egypt, Greece, or Rome, which consisted of sheets joined into a continuous roll and wound about a core, served to transmit important information of permanent value.   This preservation of important communication is probably the dominant characteristic of a "book." We could, by common agreement, decide that a "book" is any object which satisfies either the physical description or the functional characteristic we have described.   Still unanswered then would be the question of whether certain historical clay tablets could properly be classified as books.

It is not our purpose here to resolve the difficulty of sharply delineating the concept symbolized by the word "book."   Our intent is merely to call attention to this difficulty and to introduce, incidentally, two noteworthy characteristics of a symbol:

1. The particular symbol chosen to represent a concept is perfectly arbitrary.   Instead of the symbol "book" we could, by common agreement, adopt any one of a host of other symbols to represent the same concept.

2. The meaning associated with a symbol is that which we attribute to it by common consent.

The symbols used in algebra possess these same two characteristics. Thus the number symbols, such as 1, 2, 3, and so on, and the symbols for variables such as x or y are arbitrarily chosen to represent quantitative concepts.   In order to use them skillfully, or even intelligently, it is necessary to learn the meanings associated with them by common consent.

We will introduce many of the symbols of algebra in this book.   Some will deal with the relative magnitudes of quantitative expressions; these are the signs of equality and inequality.   Others, such as the signs for addition, subtraction, multiplication, and division, will deal with the operations of quantitative expressions.   Still others will be concerned with the relationship between quantitative expressions.   These will include symbols denoting set relations, functions, and variation.

The use of such symbols forms a quantitative language—the language of algebra.   When a verbal statement is translated into the language of algebra, the mathematician finds it clearer, more concise, and more useful. For example, the verbal statement

The resistance of a wire to the flow of electricity varies directly as its length and inversely as the square of its radius.

can be translated into the equivalent algebraic statement

$$R = \frac{kl}{d^2}$$

In the algebraic form this statement lends itself readily to both evaluation and analysis. As we proceed in this book, one of our interests will be to learn to understand and use algebra as a language of quantitative relationships.

Algebra as a Logical Structure

If you studied geometry, you no doubt recall that every statement in the proof of a theorem or in the solution of an "original" had to be justified by an appropriate reason. These reasons were either *definitions* (such as the term "vertical angles"), *postulates* (basic assumptions such as "quantities equal to the same quantity are equal to each other"), or *theorems* (statements which were proved, such as "the base angles of an isosceles triangle are equal"). And, of course, as in every science, there are certain terms which cannot be defined in simpler terms and therefore are accepted arbitrarily as *undefined terms*. For example, the word "point" in geometry is usually accepted as an undefined term. You may also recall that there was a very definite order in which the theorems were proved; each theorem depended on one or more theorems which were previously proved. This is what we mean by the term "logical structure."

Algebra has the same kind of logical structure as geometry. The undefined terms, defined terms, and postulates of algebra are the basis of mathematical assertions which can broadly be called conclusions. The conclusions that are of general significance are called theorems. The theorems that have been proved, along with the terms and postulates, in turn form the basis for subsequent theorems.

The proof of a theorem will usually depend on a type of logical reasoning known as *deductive reasoning*. Our development of the statement of a theorem, or our informal introduction to it, will frequently involve another type of reasoning—*inductive reasoning*. It will help us to recognize and to understand these two types of reasoning if we illustrate each with simple examples.

Inductive reasoning is a method of reasoning which establishes general laws or makes predictions on the basis of individual experiences. A visitor in a foreign country who received a fast and harrowing ride on each of the several occasions he used a taxi might conclude, by inductive reasoning, that all taxi drivers in that country drive rapidly and recklessly. A scientist who finds that mice exposed to a certain chemical irritant develop cancer might conclude, by inductive reasoning, that the irritant is carcinogenic.   Thus this type of reasoning occurs both in everyday living experiences and in the course of scientific investigations.   Indeed, inductive reasoning is probably the basis for the expression of the law of universal gravitation as a result of the observation of the mutual attraction of certain given bodies or the basis of the system which enables an insurance company to determine its risk as a result of past statistical tables, as well as the basis of the personal conviction that snow is cold or that running produces fatigue.

An important characteristic of inductive reasoning is that a single contrary example may serve to disprove the conclusion.   Consider the following:

$$1 \times 1 = 1; 1 \text{ is equal to } 1$$
$$2 \times 2 = 4; 4 \text{ is greater than } 2$$
$$3 \times 3 = 9; 9 \text{ is greater than } 3$$
$$4 \times 4 = 16; 16 \text{ is greater than } 4$$

If we continue in this manner we can develop a great many individual instances which support the conclusion that the product of a number multiplied by itself is greater than or equal to the number.   Note however that

$$\tfrac{1}{2} \times \tfrac{1}{2} = \tfrac{1}{4}; \tfrac{1}{4} \text{ is not greater than } \tfrac{1}{2}$$

This is sufficient to disprove the conclusion reached by inductive reasoning.

Deductive reasoning differs from inductive reasoning in that it proceeds from a general principle to other less general assertions which are dependent on this principle.   It is best illustrated by a *syllogism*, a logical form consisting of three assertions.   The first two are called *premises* and the third, the one to be proved, is called the *conclusion*.   For example,

If a chemical solution turns blue litmus paper red, it is an acid.
An unknown liquid turns blue litmus paper red.
Therefore the unknown liquid is an acid.

The premises may or may not be true, but if we accept the premises, and if the conclusion is a logical consequence of the premises, we must accept the conclusion.   Thus a person could assert that a storm is coming because the barometer is falling rapidly and a rapidly falling barometer portends a storm.   If in truth the barometer is falling rapidly, and if we accept the premise that this indicates a storm, we must agree with the conclusion that a storm is imminent.

Deduction differs from induction not only in that it proceeds from general laws rather than particular facts, but also in that the conclusions reached by valid deduction from indisputable premises possess a degree of certainty which can never be attained through induction alone.   We will find induction useful in discovering and developing mathematical relationships, but since mathematics is primarily deductive in nature we will be principally concerned with deductive methods.

Specifically, we will use deduction to show that every "move" you make in algebra is based on some accepted premise—a definition, postulate, or theorem.   Once you have this idea firmly in mind, not only will you understand what you are doing but why you are doing it; there will be fewer mistakes due to faulty reasoning.   When confronted by a new problem you will learn that its solution will, in most cases, be related to the solution of problems you have previously studied.

Algebra as a Method of Problem Solving

Much of mathematical development has had its roots in more or less practical requirements and the vital applications of mathematics are still one of its important features.   Algebra is used to evaluate and understand formulas, analyze relationships, and solve verbal problems—the so-called "word problems" you have encountered in your previous studies.

When the facts of a problem are translated into the language of algebra, we not only focus our efforts on all relevant details, but reduce the problem to that of routine operations.   Two skills are required:

1. The ability to translate quantitative relationships involved in a problem situation into the language of algebra.   This requires a knowledge and understanding of the use of algebraic symbols.

2. The ability to effect desired changes in the form of an algebraic statement to effect a solution.   This requires a knowledge and understanding of operations with algebraic quantities.

Through the use of these two abilities in an active experience in algebra we hope the reader will find the answer to the question, What is algebra?

CHAPTER 2

The Set of Integers

1 The Concept of Set

The concept of set appears frequently in a modern study of algebra. We agree to recognize as a set any distinct collection of objects. No attempt is made to define the word "set." Like the word "point" in geometry, it is one of the undefined words of mathematics. Other collective words, such as "class" and "collection," are sometimes used as a substitute for "set." They are considered to have equivalent meanings; behind each of these words lies the mental concept of several things which are being considered as a coherent whole. Our preference will be to use the word "set" in our discussions, although on occasion one of the other words may be used interchangeably.

The individual objects constituting a set are called *elements* of the set. One way to identify a particular set is to list all its elements. For example, the numbers 2, 4, 6, 8 form a set, and the colors red, blue, and yellow form a set.

A second way of identifying a set is to indicate the property or properties which determine whether a given object is an element of the set. Thus we can describe a set as one comprised of all mountain peaks higher than Mt. McKinley. Since Mt. Everest is higher than Mt. McKinley it is an element of this set. On the other hand, the Matterhorn is not an element of the set because its altitude is less than that of Mt. McKinley.

The Set of Natural Numbers

Of immediate interest to us is the set whose elements are the numbers used in counting, that is, the numbers 1, 2, 3, 4, . . . (the dots following a number sequence are read "and so on"). Since these numbers arise naturally in counting situations they are called *natural numbers*. They are also often referred to as *positive integers*.

A mathematical concept closely associated with the set of natural numbers is that of one-to-one correspondence. When we speak of "three" pennies, we mean that we could join with a string (either real or imaginary) each of the pennies with just one element of a set of basic counters. These counters could consist of tally marks, notches, pebbles, the fingers of our hands, or the set of natural numbers. If each element of the set of counters is paired with one of the pennies, we identify that set of counters with the number "three" or its symbol "3." Any other set which could be so joined with the set of pennies would share that identification. We say that the sets are in a one-to-one correspondence and define this concept as follows.

A *one-to-one correspondence* exists between two sets if every element of the first set can be paired with just a single element of the second set and, conversely, every element of the second set can be paired with just a single element of the first set.

For example, consider a set which consists of the sheep in a given pasture. We could form a second set consisting of tally marks such that corresponding to each sheep there is one tally mark and corresponding to each tally mark there is one sheep. The two sets are in one-to-one correspondence; there are exactly as many sheep in one set as tally marks in the other set. It is probable that primitive man used this means to convey information about the magnitude of a collection.

This technique of setting up a correspondence between elements of two sets is really the basis for our present day system of determining the extent of a collection by counting. When we count the objects in a set, we are associating each object with an element (taken in a definite order) in the set of natural numbers. We note that although there is a first natural number, 1, there is no last number; we say that the set of natural numbers is an *infinite set*. On the other hand, any set whose elements are less than a given fixed number is a *finite set* since the process of counting its elements eventually terminates.

Cardinal and Ordinal Numbers

The number associated with a set as a result of counting its elements is the same regardless of the particular order in which the elements are counted (the word "order" as used here is synonymous with the word "arrangement"). We call this number the *cardinal number* of the set. The cardinal number of a set is regarded as a property of the set; it indicates the magnitude or extent of the set. If the elements are taken in a particular order, however, we may speak of the first element, the second element, and so on. The position of any element in this arrangement is called its *ordinal number;* it is a property of the individual element of the set. Thus the cardinal number of the set of letters of our alphabet is 26; if the letters are arranged in their usual order, the ordinal number of the element *e* is 5.

Symbolic Representation of a Set

Sets are generally designated by capital letters. Thus A may be the set of all points on a line, or Q may represent the set of faculty members of a given school. The elements of a set are denoted by small letters, as a, b, c, etc., or by specific notation as Mars, poison ivy, 3, or circle. The basic relationship of an element to a set is that of belonging or being a member. This relation is denoted by the symbol $\in$, modeled on the Greek letter epsilon. Thus

$$b \in N$$

means that the object b is an element of the set N. The negation of this relation is written

$$b \notin N$$

which means that the object b is not an element of the set N.

A useful convention for symbolizing a set is to list the names of its elements and enclose them in braces, or to enclose in braces a descriptive phrase. For example, the set of all odd numbers less than 10 is

$$A = \{1, 3, 5, 7, 9\}$$

and the set of all numbers used in counting may be denoted as

$$N = \{\text{natural numbers}\}$$

The letter x is frequently used to represent any element of a given set. Thus if x is any natural number, we can specify a particular set as

$$K = \{x, \text{ such that } x \text{ is greater than 5}\} \tag{1}$$

The set then consists of the numbers 6, 7, 8, 9, and so on. Used in this way, x is called a *variable*. It is a symbol for any one of a specified set of elements.

Statement (1) can also be written in the form

$$K = \{x \mid x > 5\}$$

In this notation the vertical bar is read "such that," and the symbol ">" is read "greater than," and the entire set notation is read "K is the set of all x's such that x is greater than 5."

Since signs of equality and inequality may be used in a variety of ways to indicate restrictions on the variable x, we list some common forms of these symbols together with their meanings:

$=$ equals	$\geq$ greater than or equal to
$\neq$ does not equal	$\leq$ less than or equal to
$>$ greater than	$\not> $ is not greater than
$<$ less than	$\not< $ is not less than

As a further illustration of this symbolism consider the notation

$$B = \{x \in U \mid x \neq 3\} \quad \text{where} \quad U = \{1, 2, 3, 4\}$$

This means that set B consists of all elements of U which are not equal to 3. Therefore

$$B = \{1, 2, 4\}$$

Exercise 1

THE CONCEPT OF SET

List the elements of the following sets:

1. All natural numbers less than 30 which are divisible by 3.

2. All months of the year which have 31 days.

3. All months of the year which have more than 31 days. (This is an example of a null, or empty set.)

4. All pairs of natural numbers whose sum is 7.

5. All equilateral triangles with one right angle.

Which of the following numbers do not belong to the set of natural numbers?

6. 33	**7.** $3\frac{1}{2}$	**8.** 0	**9.** $-\frac{3}{4}$	**10.** $\frac{3}{4}$	**11.** 251
12. 4.5	**13.** $\frac{4}{3}$	**14.** -2	**15.** 3.1416	**16.** 0.5	**17.** 13

Which of the following are elements of the set of all linear units of measurement?

18. inch	19. meter	20. foot
21. gram	22. kilometer	23. kilogram
24. yard	25. cubic inch	26. acre
27. rod	28. mile	29. radian

30. Name the elements of the set of all regular polygons which can be used as floor tiles.

Which of the following pairs of sets have a one-to-one correspondence?

31. State and state capitals.   32. Words and groups of letters.

33. Bridges and rivers.   34. Fingers and toes.

35. Odd numbers from 1 to 99   36. Wheels and tires of an auto-
and even numbers from 2 to 198.   mobile.

Give the cardinal number of each of the following sets:

37. Letters of the English alphabet.
38. Letters of the Greek alphabet.
39. Months of the year
40. Cards in a heart suit.
41. Days of February, 1967.
42. Presidents of the United States who have served two or more terms by election.

Give the ordinal number of each of the following stated elements. Assume the elements to be in their natural order.

43. The letter h, in the set specified in Problem 37.
44. The letter π, in the set specified in Problem 38.
45. April, in the set specified in Problem 39.
46. The jack of hearts, in the set specified in Problem 40.
47. St. Valentine's Day, in the set specified in Problem 41.
48. Franklin D. Roosevelt, in the set specified in Problem 42.
49. It has been said that trying to define every word is like trying to learn Russian from a Russian dictionary.*   Explain this statement.
50. Give some examples of numbers used in everyday life which are not elements of the set of natural numbers.

* R. B. Kershner and L. R. Wilcox, *The Anatomy of Mathematics*, The Ronald Press Company, New York, 1950; page 10.

If a and b are elements of the set of natural numbers S, indicate which of the following statements are true and which are false:

51. $a + b$ is an element of S.
52. $a - b$ is an element of S.
53. $a - b$ may be an element of S.
54. $a \times b$ is an element of S.
55. a/b cannot be an element of S.

Indicate whether each of the following is a finite or an infinite set:

56. The set of all grains of sand on all beaches at a particular moment.
57. The set of all colors.
58. The set of points on a given line.
59. The set of all fractions less than 1.
60. The set of all odd numbers.

Write a description for each of the following sets of numbers, as illustrated in Problem 61:

61. $\{11, 13, 15, 17, 19\}$—The set of all odd numbers between 10 and 20.
62. $\{4, 8, 12, 16, 20\}$
63. $\{36, 49, 64, 81\}$
64. $\{1, 8, 27, 64, 125\}$
65. $\{6, 16, 26, 36\}$

For each of the following, specify a set whose elements are:

66. Educational institutions.
67. Plane geometric figures.
68. Geographic locations.
69. Insects.
70. Chemical compounds.

If $U = \{2, 4, 6, 8, 10\}$, list the elements of each of the following sets:

71. $\{x \in U \mid x \neq 4\}$ **72.** $\{x \in U \mid x > 6\}$
73. $\{x \in U \mid x < 10\}$ **74.** $\{x \in U \mid x \geq 8\}$
75. $\{x \in U \mid x \not> 4\}$ **76.** $\{x \in U \mid x \not< 4\}$
77. $\{x \in U \mid x = 6\}$ **78.** $\{x \in U \mid x \leq 10\}$
79. $\{x \in U \mid 2 < x < 10\}$ **80.** $\{x \in U \mid 4 < x \leq 10\}$

2 The Fundamental Operations

The fundamental operations of algebra are addition, multiplication, subtraction, and division. These operations are closely related to the process of counting. They are also related to each other; the fact that subtraction may be checked by addition and that division may be checked by multiplication is a common use of this relationship. In the following discussion some statements will be given in terms of variables such as *a*, *b*, and *c* to give them greater generality. It will be our understanding that we can replace each of these letters by any element of the set of natural numbers.

Addition of Natural Numbers

If a man has 8 dollars in his wallet and receives 5 more dollars, he may recount all of the bills to find the total. He may also find the total by the operation of addition, for he has memorized the fact that 5 added to 8 equals 13. This fact, stated in the language of arithmetic, becomes $8 + 5 = 13$. The result, 13, is called the *sum* and the numbers added together are called *terms* of the sum. The symbol "$+$" indicates the operation of addition. To be more general, if a set of elements whose cardinal number is *b* is added to a set of like elements whose cardinal number is *a*, then $a + b = c$, where *c* is the cardinal number of the combined sets. We note that the addition of any two natural numbers is always possible, and that the sum is always a natural number. We have not actually defined the term addition; it remains as yet an undefined term.

Multiplication of Natural Numbers

If 6 bars of soap costing 9 cents a bar are purchased, the cash register tape shows this purchase as 9, 9, 9, 9, 9, 9; the total price is 54 cents, the addition being performed by the cash register. The purchaser, however, would be more likely to figure the cost as $6 \times 9 = 54$. In the first case, 54 is the sum of six 9's; in the second case, 54 is the product of 6 times 9. We are thus led to the definition of multiplication as *repeated addition*. It should be borne in mind that the shorter process of multiplication is based on memorization of the multiplication tables in the same way that addition is based on memorization of certain number combinations.   There are several ways of indicating the operation of multiplication: 6×9, $6(9)$, $(6)(9)$, and $6 \cdot 9$, and all mean that 9 is multiplied by 6.   Since the use of the symbol $\times$ for multiplication may cause confusion when we are using letters, we write ab to mean b multiplied by a.   In the multiplication $ab = c$, a and b are called *factors*, and c is the *product*.   We note that the multiplication of two natural numbers is always possible, and that the product of any two natural numbers is a natural number.

Subtraction of Natural Numbers

The 6 bars of soap referred to in the last section cost 54 cents (we assume no sales tax).   What change will the customer receive if he hands the clerk 75 cents?   We may say that 54 subtracted from 75 equals 21, so the change is 21 cents.   But when the change is handed to the customer, in most cases the clerk adds to 54 an amount necessary to obtain 75 as a check on his subtraction.   But this addition is actually more than a check, for the operation of subtracting 54 from 75 implies the problem of addition: what number added to 54 gives 75?

We are thus led to the realization that subtraction is closely related to addition, in fact, we could express the ideas involved without introducing the new operation.   It is neither necessary nor desirable, however, to deprive ourselves of its use.   In general the problem is to find a number which added to a given number produces a given sum.   If a is a given number and b is a given sum, then we are seeking a number c such that $a + c = b$.   We define the process of finding one term when the sum and the other term are given as *subtraction*.   We shall use the symbol "$-$" in the usual sense to express this relationship, that is, if $a + c = b$, then $c = b - a$.   The result of the process is called the *difference* and the numbers b and a are called the *terms* of the difference.   We note that subtraction of any two natural numbers is not always possible; there is no natural number c such that $5 - 7 = c$.

Subtraction and addition are called *inverse* operations.   The use of the term inverse may be justified as follows.   If we begin with the number a and add to it the number b, then subtract the number b from the sum, we return to the initial number a.   Thus the operation of subtraction undoes the operation of addition and in this sense is said to be the inverse of addition.   Any two operations so related that one countereffects the other are said to be inverse operations.

Division of Natural Numbers

If we wish to divide 15 dollars equally among 5 people we may think, "5 times what number equals 15?" Since 5 times 3 equals 15, we can say that 15 divided by 5 equals 3. In this sense division is the inverse of multiplication. We may also think of division as repeated subtraction, for as each person receives 3 dollars, 3 is subtracted from what is left until no dollars remain. Since repeated subtraction is the inverse of repeated addition, that is, of multiplication, both considerations lead us to the definition of division as the *inverse of multiplication*. To be more general, using the symbol "$\div$" to denote division, $a \div b = c$ if $bc = a$. In the division $a \div b = c$, a is the *dividend*, b is the *divisor*, and c is the *quotient*. We note that the division of two natural numbers is not always possible; there is no natural number c such that $16 \div 5 = c$.

Powers and Roots of Natural Numbers

The process of raising a number to a power and the process of extracting a root of a number, although not defined as fundamental operations, are closely related to them. If we wish to find the area of a square with a 6 inch side, we may state that the area equals $6 \times 6 = 36$, the unit being square inches. A more convenient way of expressing the product 6×6 is by using the exponent 2, that is, 6×6 may be written as 6^2 (read "six squared," or "six to the second power," or "the second power of six"). Similarly, $6^3 = 6 \times 6 \times 6 = 216$, $6^4 = 6 \times 6 \times 6 \times 6 = 1296$, and so on. More generally, $b^n = b \cdot b \cdot b \cdots b$ to n factors, n also being a natural number. In the expression $b^n = c$, b is the *base*, n is the *exponent*, and c is the *nth power of b*. We define this operation as *repeated multiplication*, and we note that raising a natural number to a power (which itself is a natural number) always results in a natural number.

If we wish to find the two equal factors of 36, we indicate this by the expression $\sqrt{36} = 6$, since $6 \times 6 = 36$. Similarly $\sqrt[3]{27} = 3$, since $3 \times 3 \times 3 = 27$. More generally, $\sqrt[n]{N} = b$, if $b \cdot b \cdot b \cdots b$ to n factors equals N, where N, n, and b are natural numbers. In the expression $\sqrt[n]{N} = b$, N is the *radicand*, n is the *index*, and b is the *root*. The symbol $\sqrt{}$ is called the *radical sign*. The expression $\sqrt{36}$ is read "the square root of 36," the expression $\sqrt[3]{27}$ is read "the cube root of 27," and $\sqrt[4]{16}$ is read "the fourth root of 16." We note that finding a given root of any natural number is not always possible in the set of natural numbers; there is no natural number b such that $\sqrt{5} = b$.

Exercise 2

THE FUNDAMENTAL OPERATIONS

Which of the fundamental operations are implied in each of the following statements?   Do not solve the problem.

1. The total weight of 100 bolts weighing 1 ounce each.
2. The cost of one bar of candy if a carton of 12 costs 60 cents.
3. The cost of 12 different items purchased at a food store.
4. The cost of mailing 100 post cards.
5. The net weight of a box of soap, given the gross weight and the weight of the box.
6. The cost resulting from using a coupon worth 10 cents towards the purchase of an article.
7. The perimeter of a square 2 inches on a side.

State which of the following problems require finding the power of a number and which require finding the root of a number:

8. The area of a square 2 inches on a side.
9. The side of a square whose area is 100 square inches.
10. The volume of a cube 3 inches on a side.
11. The side of a cube whose volume is 125 cubic inches.
12. The number of cubic inches in a cubic foot.
13. The number of square inches in a square foot.
14. The number of toy building blocks along one edge of a cubical pile of 343 blocks.

Which of the following operations are impossible in the set of natural numbers?

15. $7 \div 3$	16. $\sqrt[3]{125}$	17. $\sqrt{75}$	18. $8 - 8$	19. 8×8
20. $8 \div 8$	21. $8 + 8$	22. 8^3	23. $\sqrt[3]{8}$	24. $3 \div 7$
25. $7 - 9$	26. $9 - 7$	27. 4^3	28. 3^4	29. 6^0
30. 7×0	31. $17 - 0$	32. $7 - 0$	33. $7 + 0$	34. $0 - 7$

In each of the following problems, state whether 6 is a sum, a difference, an exponent, a product, a factor, a root, a base, or an index:

35. $\sqrt{36} = 6$ **36.** $(12)(6) = 72$ **37.** $6^2 = 36$ **38.** $\sqrt[6]{64} = 2$
39. $2^6 = 64$ **40.** $2 + 4 = 6$ **41.** $8 - 2 = 6$ **42.** $(3)(2) = 6$
43. Show by repeated subtraction that 24 divided by 4 equals 6.
44. Show by repeated subtraction that when 24 is divided by 5, the remainder is 4.

Find the following powers of natural numbers:

45. 2^2 **46.** 3^2 **47.** 4^2 **48.** 5^2 **49.** 6^2 **50.** 2^3 **51.** 3^3 **52.** 4^3
53. 5^3 **54.** 6^3 **55.** 2^4 **56.** 3^4 **57.** 4^4 **58.** 5^4 **59.** 6^4

Find by trial the roots of the following natural numbers:

60. $\sqrt{9}$ **61.** $\sqrt{4}$ **62.** $\sqrt{25}$ **63.** $\sqrt[3]{8}$ **64.** $\sqrt[3]{125}$
65. $\sqrt[3]{27}$ **66.** $\sqrt[4]{625}$ **67.** $\sqrt[4]{1296}$ **68.** $\sqrt[3]{216}$ **69.** $\sqrt[4]{81}$
70. $\sqrt{36}$ **71.** $\sqrt{16}$ **72.** $\sqrt[3]{64}$ **73.** $\sqrt[4]{256}$ **74.** $\sqrt[4]{16}$

Rewrite each of the following as an indicated power or product. Do not evaluate the result.

75. $3 + 3 + 3 + 3$
76. $3 \times 3 \times 3 \times 3$
77. $5 + 5 + 5$
78. $7 \times 7 \times 7 \times 7 \times 7$
79. $13 \times 13 \times 13 \times 13 \times 13 \times 13$
80. How many times can 8 be subtracted from 56? What single operation is equivalent to this repeated subtraction?

In the following, p and q are symbols for elements in the set of natural numbers.

81. If $p + 4 = q$, write two relations of these numbers using the operation of subtraction.
82. If $7 \times p = q$, write two relations of these numbers using the operation of division.
83. If $5 \times p = q$, write the relation of p and q as a repeated addition.

Given the numbers 6 and 4, find:

84. The sum of their squares.
85. The square of their sum.

86. The difference of their squares.
87. The square of their difference.

3 Working with Postulates

In discussing the fundamental operations with natural numbers, we noted that the operations of addition and multiplication were always possible.   We are now ready to state this fact and certain other laws governing the fundamental operations in the form of postulates.   This is our introduction to algebra as a logical structure.

Although our initial reference will be to the set of natural numbers, it is important to keep in mind that these numbers do not constitute the only set which satisfies our postulates.   As we consider more extensive number sets it will become apparent that the postulates are shared by many other sets of numbers, and mathematicians have therefore found it advantageous to study them from "the postulational point of view." Most students will recognize this approach as similar to that used in the study of geometry.   Seven postulates are considered at this point and four more are introduced shortly to form a set of eleven postulates upon which we base our development of an algebraic system in which one can add, subtract, multiply, and divide.

In the statement of our assumptions we again use the variables a, b, and c to denote elements of the set of natural numbers.   We also use the parenthesis as a grouping symbol to indicate that whatever is written inside is to be understood as one quantity.   For example, $7(3 + 2)$ indicates the product of 7 and the sum of 3 and 2, that is,

$$7(3 + 2) = 7(5) = 35$$

Similarly the expression $(a + b) + c$ indicates the addition of the sum $a + b$ and c.

The Postulates for the Addition of Natural Numbers

Postulate 1.   *The sum of any two natural numbers, a and b, in the stated order, is a unique natural number c.*

This property is known as the *closure law* of addition.   The term "closure" is used to imply that the set of natural numbers is closed with respect to the operation of addition, that is, the sum of any two elements of the set must be an element of the set. The closure law states also that this sum will be unique; there is one and only one element which is the sum of two given elements.

Postulate 2. *If a and b are any two natural numbers, then* $a + b = b + a.$

This property is known as the *commutative law* of addition. The closure postulate asserted that the sum exists and is unique; the commutative postulate asserts that this sum is independent of the sequential order in which the two elements are added. For example, $2 + 7 = 7 + 2$. This property and each of the other properties listed in this section are of course very familiar to the reader as a result of years of experience with such techniques. That this property is not intrinsically obvious is apparent if we recall that the growing child may have to color many pictures before he discovers that the result of counting "2 more than 7" is the same as the result of counting "7 more than 2."

Postulate 3. *If a, b, and c are any three natural numbers, then* $(a + b) + c = a + (b + c).$

This property is known as the *associative law* of addition. Since the first two postulates dealt with the sum of *two* numbers, an expression of the form $a + b + c$ is not meaningful. If, however, we interpret $a + b + c$ to mean $(a + b) + c$, the addition does have meaning because, by Postulate 1, the unique sum $a + b$ exists and we can then add this sum to the number c. It is also meaningful to interpret $a + b + c$ as $a + (b + c)$ using the same line of reasoning. The associative postulate now asserts that these two interpretations are equivalent: the sum of the first two numbers may be added to the third number or the first number may be added to the sum of the last two numbers. For example, the sum $3 + 4 + 5$ equals $7 + 5$ or $3 + 9$; in both cases the sum is 12.

We now define $a + b + c$ to mean $(a + b) + c$ and this, by the associative law, is equal to $a + (b + c)$. In a similar manner we can define the sum of four elements of the set, then five elements, and so on.

EXAMPLE 1. Prove $(c + b) + a = (a + b) + c.$

Proof. $\begin{aligned}(c + b) + a &= c + (b + a) &&\text{(associative law)}\\ &= (b + a) + c &&\text{(closure law;}\\ &&&\text{commutative law)}\\ &= (a + b) + c &&\text{(commutative law)}\end{aligned}$

Note. In the second step of the proof, the closure law is necessary to interpret the quantity $(b + a)$ as a single element. The commutative law then justifies the interchange of sequence of the two elements.

The Postulates for Multiplication of Natural Numbers

Postulate 4. *The product of any two natural numbers, a and b, in the stated order, is a unique natural number c.*

Postulate 5. *If a and b are any two natural numbers, then $ab = ba$.*

Postulate 6. *If a, b, and c are any three natural numbers, then $(ab)c = a(bc)$.*

The laws of multiplication are analogous to those of addition, with "product" replacing "sum." Postulates 4, 5, and 6 state, respectively, the closure law, commutative law, and associative law for natural numbers with respect to multiplication.

EXAMPLE 2. Prove $(ab)c + d = d + a(cb)$.

Proof. $(ab)c + d = d + (ab)c$ (commutative law of addition)

$\qquad\qquad\quad = d + a(bc)$ (associative law of multiplication)

$\qquad\qquad\quad = d + a(cb)$ (commutative law of multiplication)

The Distributive Postulate

One more postulate is important in working with natural numbers; it connects the operations of addition and multiplication.

Postulate 7. *If a, b, and c are any three natural numbers, then $a(b + c) = ab + ac$.*

This property is known as the *distributive law.* It states that multiplication is distributive with respect to addition. For example, $7(2 + 4) = 7 \cdot 2 + 7 \cdot 4 = 14 + 28 = 42$. If we write $7(2 + 4)$ as $7 \cdot 6$, the product is again 42.

The distributive law, read from left to right, changes a product of the form $a(b + c)$ into the sum of ab and ac. We refer to this change as "multiplying out" the expression. Reading this postulate from right to left changes a sum of two products having a common factor into a product of the form $a(b + c)$. We refer to this change as "taking out a common factor." For example, note that the formula for the perimeter of a rectangle may be written as $P = 2l + 2w$ or $P = 2(l + w)$.

Exercise 3

Working with Postulates

1. Show by several examples that the commutative and associative laws do not apply to the operations of subtraction and division of natural numbers.

Identify the postulate which is the basis for each of the following statements or operations (all letters represent natural numbers):

2. $2 \cdot 7 = 7 \cdot 2$ *com*   **3.** $ac = ca$ **4.** $a(x + y) = ax + ay$
5. $3 + (4 + 7) = (3 + 4) + 7$ **6.** $6 + 7 = 7 + 6$ **7.** $(xy)z = x(yz)$
8. $ab + ac = a(b + c)$ **9.** $a + (b + c) + d = a + b + (c + d)$

Prove that each of the following equalities are true by using one or more of the postulates stated in this section:

10. $(ab)c = c(ab)$ **11.** $a(bc) = c(ab)$ **12.** $(ab)c = c(ba)$
13. $a + (b + c) = c + (a + b)$
14. $(a + b) + c = a + (c + b)$
15. $(3x + 3y)z = 3z(x + y)$
16. $(a + 2)(b + c) = (a + 2)b + (a + 2)c$
17. $(a + 2)(b + c) = ab + 2b + ac + 2c$

A set is closed under a given operation if all the elements obtained by applying the operation remain within the set. For example, the set of even numbers is closed under the operation of multiplying by 2 since all elements so obtained will be even numbers.

State which of the following sets are closed under the given operation:

Set	Operation
18. All even numbers	Multiply by 3
19. All even numbers	Multiply by any natural number
20. All even numbers	Divide by 2
21. All odd numbers	Multiply by 2
22. All odd numbers	Multiply by 3
23. All perfect squares	Multiply by 2
24. All perfect squares	Multiply by 4
25. All words	Changing order of the letters

Which of the following operations are commutative?

26. Taking a shower and getting dressed.
27. Pouring acid into water.
28. Mowing the lawn and trimming the hedge.
29. Mowing the lawn and raking the lawn.
30. Subtracting 5 from 10.
31. Dividing 25 by 5.
32. Cleaning a car and then waxing it.
33. Let aRb represent the assertion "a is related to b."

   (a) If a and b are people, list two examples in which $aRb = bRa$ and two examples in which $aRb \neq bRa$.

   (b) If a and b are countries, list two examples in which $aRb = bRa$ and two examples in which $aRb \neq bRa$.

4   Interpretation of Directed Numbers and Zero

Extending the Number System

We have carefully pointed out the fact that subtraction is not always possible in the set of natural numbers.   In fact, if we limit ourselves to the use of natural numbers, it is not possible to give a numerical answer to the simple problem: "A man has \$10 and spends \$10.   What does he have left?"   The answer is, of course, \$0, but zero is not a natural number.   In fact, confining ourselves to the set of natural numbers, it is not possible to state the temperature on a certain day when there was a drop of 10° from a previous reading of 7°, nor is it possible to state the centigrade equivalent of 32°F.   To give meaning to such operations, we extend the number system to include zero and the negative numbers.

Definition of Zero

The natural numbers have been defined as the numbers used in counting.   Since zero is not used in counting, we introduce zero into our number system as follows.

Postulate 8.   *For any element a of the set of natural numbers, there exists an element I such that $a + I = I + a = a$.*

The number I defined by this postulate is considered as the *identity element* for the operation of addition within the set under consideration.   It is commonly called zero and is denoted by the symbol 0. The set of numbers consisting of the natural numbers and zero is called the set of *whole numbers*.

Therefore, by definition, zero obeys the closure postulate for addition, since $a + 0 = a$. Also since $a \cdot 0 = 0 + 0 + 0 + \cdots + 0$ (a times), it is clear that $a \cdot 0 = 0$, and therefore zero obeys the closure postulate for multiplication. Similarly, the number zero obeys each of the other postulates we have stated for natural numbers:

$a + 0 = 0 + a$	(commutative law of addition)
$a \cdot 0 = 0 \cdot a$	(commutative law of multiplication)
$(a + b) + 0 = a + (b + 0)$	(associative law of addition)
$(a \cdot b) \cdot 0 = a \cdot (b \cdot 0)$	(associative law of multiplication)
$a(b + 0) = ab + a \cdot 0$	(distributive law)

Directed Numbers

If we indicate a temperature of $3°$ below zero as $-3°$, there is little chance of confusion. We simply mean that the temperature is $3°$ lower than a temperature which we arbitrarily call $0°$ (of course we must specify whether the temperature is measured in the Fahrenheit or centigrade scale). But if we speak of -5 books, this statement has no meaning in the physical sense; we cannot point to a set of -5 books. Referring to the temperature of $-3°$ again, we see that the idea of direction is involved; a temperature of $+3°$ is $3°$ above zero and a temperature of $-3°$ is $3°$ below zero. It is therefore convenient to think of $+3$ and -3 as directed numbers.

We have thus introduced an important concept. Corresponding to each natural number a there is a positive integer which we shall designate as $+a$ and a negative integer which we shall designate as $-a$. For our purposes it will not be necessary to distinquish between the natural number a and the positive integer $+a$; we define $+a = a$. The negative integer $-a$, called the *negative of a*, is defined by the following postulate.

Postulate 9. *For every element a of the set of natural numbers, there exists an element $-a$ such that $a + (-a) = (-a) + a = 0$.*

Each of the two elements a and $-a$ is said to be the additive inverse of the other. Thus -3 is the additive inverse of $+3$ and 5 is the additive inverse of -5.

The number system has now been extended to contain positive integers (natural numbers), negative integers, and zero.   This collection of numbers is called the *set of integers*.   The arithmetic behavior of this set of numbers is in agreement with the postulates and definitions previously listed.   We shall refer to these numbers, and certain other numbers which will be discussed later (sections 9 and 26), as *directed numbers*.   The purpose of this designation is to emphasize, at appropriate times, the sense of "direction" implicit in the sign associated with each number. These numbers are also referred to frequently as *signed numbers*.

The Number Scale

It is convenient to represent directed numbers and zero by means of a number scale, as in Figure 1.   Concerning this number scale, we note that:

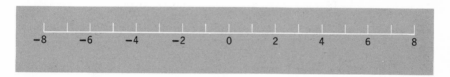

Figure 1

1. The positive direction is to the right of zero.
2. The negative direction is to the left of zero.
3. A number on the scale is greater than any number to the left of it. In symbols, $3 > 1$; $-1 > -2$.
4. A number on the scale is less than any number to the right of it. In symbols, $-4 < 2$; $-4 < -1$.

The Absolute Value

A remark about distances measured along a number scale may clarify a point of frequent confusion.   If A travels 3 miles to the right of 0, then he reaches a point $+3$ miles.   If B travels 4 miles to the left of 0, then he reaches a point -4 miles.   We agree that -4 is less than $+3$.   However, this does not imply that B traveled a lesser distance than A; it merely means that B is to the left of A on the number scale.   We say that the *absolute value* of the distance -4 is 4 miles, and the absolute value of the distance $+3$ is 3 miles.

The absolute value of a number is therefore the number of units distance corresponding to a signed number, without regard to the direction in which the distance is measured.   Absolute value is indicated by writing the signed number between vertical bars.   Thus

$$|-4| = 4, \quad |+3| = 3, \quad |0| = 0$$

We note that the absolute value of a positive integer is that integer itself, the absolute value of 0 is 0, and the absolute value of a negative integer is its additive inverse.   A formal definition is given by the following relations:

$$|x| = \begin{cases} x & \text{if } x \geq 0 \\ -x & \text{if } x < 0 \end{cases}$$

multiples of

Exercise 4

INTERPRETATION OF DIRECTED NUMBERS

On the number scale, A, B, C, D, etc., represent directed numbers on the scale.

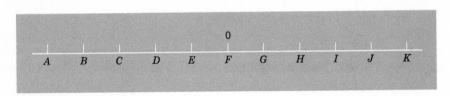

Figure 2

Place the proper sign, $>$ or $<$, between each pair of given numbers (for example, $C < G$):

1. E B	**2.** H J	**3.** G E	**4.** G H
5. I J	**6.** J I	**7.** D F	**8.** A C

Which of the following quantities might be measured by directed numbers?

9. The score in a baseball game.

10. The score in a game of shuffleboard.

11. The hourly changes in temperature.

12. The hourly temperatures in New York City.

13. The year of an event after the birth of Christ.

14. The year of an event before the birth of Christ.

15. The measurements of a land survey.

16. The errors in the measurements of a land survey.

17. The dimensions of a machine part.

18. The tolerances in the dimensions of a machine part.

If a gain of $10 is represented by $+10$, state the directed number which will represent:

19. A profit of $3.

20. A loss of $2.

21. A bank deposit of $50.

22. A bank withdrawal of $8.

23. A $200 increase in the cost of an automobile.

24. A $100 discount in the purchase of a large freezer.

If a rise in temperature of ten degrees is represented by $10°$, state the directed number which will represent:

25. A rise of $3°$.

26. A drop of $2°$.

27. An increase of $7°$.

28. A decrease of $5°$.

Find the value of:

29. $7 + |-3|$ **30.** $7 + |+3|$ **31.** $12 + |-2|$ **32.** $12 + |+2|$

33. $6 + |-3| + |-2|$ **34.** $3 + |-17| + |+2|$ **35.** $3 + |-2| + |+2|$

36. $6 \times |-3|$ **37.** $8 \times |-4|$ **38.** $\dfrac{4}{|-2|}$

39. $\dfrac{8}{|-2|}$ **40.** $12 - |-2|$ **41.** $8 - |-3|$

5 Addition, Multiplication, and Powers of Directed Numbers

Having extended the number system to include negative integers and zero, we now consider the fundamental operations with these numbers. The rules for operations with directed numbers will be formulated in such a way that the assumptions stated in the original postulates for natural numbers will apply to the larger set.   We do not prove this, but our examples and problems will verify that this is so.

Addition of Directed Numbers

The game of shuffleboard is probably familiar to most readers. The possible number of points a player may get in one trial is 7, 8, 10, or −10, as indicated on the typical shuffleboard court. We may think of the

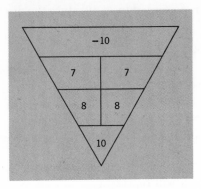

Figure 1

−10 space as meaning that 10 points are to be deducted from the player's score. But since we consider the final score as being the sum of all the points earned, we may think of the −10 space as meaning that −10 is to be added to the player's score in the same way that 7, 8, or 10 is added.

Suppose a player with a score of 7 scores −10 on his next try. His score is then $+7 + (-10)$ or −3. If he then scores 8 points, his score is $-3 + (+8)$ or +5. If he gains 10 points on his next trial, his score becomes $+5 + (+10)$ or +15. On the other hand, suppose that when his score was −3 he again scores −10. His score is then $-3 + (-10)$ or −13. We may indicate these examples as the four addition problems:

$$
\begin{array}{cccc}
+7 & -3 & +5 & -3 \\
-10 & +8 & +10 & -10 \\
\hline
-3 & +5 & +15 & -13
\end{array}
$$

Let us now consider more formally the process of adding two directed numbers. For this purpose we will again refer to the number scale introduced in Section 4. We make the following agreement:

To add a positive number a to any number b is to find the number that occupies the ath place after (to the right of) b on the number scale. To add a negative number $-a$ to any number b is to find the number that occupies the ath place before (to the left of) b on the number scale.

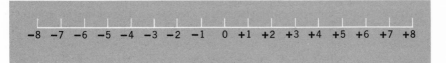

Figure 2

Thus to add $+3$ to -5, we count three places to the right of -5, so that $(-5) + (+3) = -2$. To add -3 to -5, we count three places to the left of -5, so that $(-5) + (-3) = -8$. Let us find the four sums:

(a)	(b)	(c)	(d)
$+5$	-5	$+5$	-5
$+3$	-3	-3	$+3$

(a) Starting at $+5$, we count three places to the right, so that $(+5) + (+3) = +8$.

(b) Starting at -5, we count three places to the left, so that $(-5) + (-3) = -8$.

(c) Starting at $+5$, we count three places to the left, so that $(+5) + (-3) = +2$.

(d) Starting at -5, we count three places to the right, so that $(-5) + (+3) = -2$.

By generalizing from these examples we can state the following rules for the addition of directed numbers:

1. To add two numbers with like signs, find the sum of their absolute values and give this sum the common sign.

2. To add two numbers with unlike signs, find the difference of their absolute values and give this difference the sign of the number having the larger absolute value.

Thus in addition

$+8$	-8	$+3$	-8	-4	-3	$+3$	$+6$	-3
$+3$	$+3$	-8	-3	-7	$+7$	-7	-6	$+3$
$+11$	-5	-5	-11	-11	$+4$	-4	0	0

Addition of More than Two Directed Numbers

To add more than two directed numbers, we may add the positive and the negative numbers separately, and then add the positive total and the negative total for the complete sum. We may also add the numbers in the order given, observing the proper sign at each step. Thus to add

$$(+8) + (-3) + (+2) + (-4) + (+6) + (-8)$$

we have

$$(+8) + (+2) + (+6) = +16$$
$$(-3) + (-4) + (-8) = \underline{-15}$$
$$+1 \text{ the sum}$$

We may also have

$$(+8) + (-3) = +5$$
$$(+5) + (+2) = +7$$
$$(+7) + (-4) = +3$$
$$(+3) + (+6) = +9$$
$$(+9) + (-8) = +1, \text{ as before}$$

Multiplication of Directed Numbers

If we wish to extend to directed numbers the concept of multiplication as repeated addition, the product $(+5)(-6)$ equals

$$(-6) + (-6) + (-6) + (-6) + (-6) = -30$$

Generally, we define the product $(+a)(-b)$ to be $-ab$. Since the commutative law of multiplication is to hold for directed numbers, we state the definition in the form

$$(+a)(-b) = (-b)(+a) = -ab \tag{1}$$

What value shall we assign to $(-a)(-b)$, that is, the product of two negative numbers? Consider the product $(-5)(-6)$. We could reasonably define this product to be $+30$ or -30; the postulates of closure, commutativity, and associativity for multiplication could be satisfied by either definition. However, the distributive postulate demands that we define the product to be $+30$. To verify this, consider the product $(-5)[(+6) + (-6)]$, which has the value zero. Applying the distributive law we have

$$(-5)(+6) + (-5)(-6) = 0$$

By our preceding definition, $(-5)(+6) = -30$, so we have

$$(-30) + (-5)(-6) = 0$$

Now to avoid contradiction of our definition of negative numbers we find it necessary to choose that $(-5)(-6) = +30$.

In a similar manner, if we apply the distributive law to the product $(-a)[(+b) + (-b)]$ we are led to the necessary definition:

$$(-a)(-b) = +ab \tag{2}$$

From previous considerations we also have

$$(+a)(+b) = +ab \tag{3}$$

Statements 1, 2, and 3 are equivalent to the following rule for the multiplication of directed numbers:

> The product of two directed numbers having opposite signs is negative and the product of two directed numbers having like signs is positive.

Illustrating this rule, we have

$$(-5)(+7) = -35$$
$$(-7)(+5) = -35$$
$$(-7)(-5) = +35$$
$$(+7)(+5) = +35$$

If there are more than two numbers forming the product, they can be multiplied two at a time by repeated use of the closure and associative laws until the final result is obtained.   Thus

$$(+2)(-3)(+6) = (-6)(+6) = -36$$

and

$$(-5)(+3)(-1)(+2) = (-15)(-1)(+2) = (+15)(+2) = +30$$

It will be apparent that the product is positive if there is an even number of negative factors and the product is negative if there is an odd number of negative factors.   Perhaps the rules for multiplying directed numbers may appear somewhat artificial and unrelated to the practical use of numbers.   Problem 73 of Exercise 5 is one illustration that the rules of multiplication of directed numbers lead to results compatible with the physical interpretation of the problem.

Powers of Directed Numbers

The power of a directed number may be found by applying the preceding principles of multiplication of directed numbers. This follows directly from the definition of the nth power of b:

$$b^n = b \cdot b \cdot b \cdots b \text{ (to } n \text{ factors, } n \text{ being a natural number)}$$

Thus

$$(-3)^3 = (-3)(-3)(-3) = -27$$
$$(-3)^4 = (-3)(-3)(-3)(-3) = +81$$

Exercise 5

ADDITION, MULTIPLICATION, AND POWERS OF DIRECTED NUMBERS

Find the following sums by using a number scale:

1. $(+5) + (+8)$	**2.** $(-3) + (-7)$	**3.** $(+2) + (-3)$
4. $(-8) + (-2)$	**5.** $(-8) + (+2)$	**6.** $(+9) + (-1)$
7. $(+7) + (-7)$	**8.** $(-3) + (+3)$	**9.** $(-6) + (+6)$
10. $(+2) + (-2)$	**11.** $(-3) + (-3)$	**12.** $(+4) + (+11)$
13. $(+8) + (-8)$	**14.** $(-3) + (-2)$	**15.** $(+2) + (-6)$
16. $(-2) + (-6)$		

Find the following sums by use of the number scale:

17. (a) $(+5) + (-3)$ **18.** (a) $(-3) + (-6)$ **19.** (a) $(+8) + (-2)$
(b) $(-3) + (+5)$ (b) $(-6) + (-3)$ (b) $(-2) + (+8)$

20. What law of addition is illustrated by the results of Problems 17–19?

Find the following sums without the use of a number scale:

21. $+7$	**22.** -7	**23.** -3	**24.** $+3$	**25.** -7	**26.** -3
-3	$+2$	-4	-2	$+4$	-3

27. -6	**28.** -6	**29.** $+45$	**30.** -63	**31.** -72	**32.** $+34$
-6	$+6$	-33	$+27$	$+64$	-28

33. $+28$	**34.** -28	**35.** -28	**36.** -17	**37.** -19	**38.** -56
$+32$	-32	$+17$	$+28$	$+19$	$+56$

Find the following sums:

39. $(+7) + (+6) + (-2)$ **40.** $(+7) + (-6) + (-3)$
41. $(-17) + (-2) + (+4)$ **42.** $(-32) + (+17) + (-2)$
43. $(-2) + (+18) + (-37)$ **44.** $(-3) + (-4) + (-2) + (-9)$
45. $(-13) + (-14) + (+27) + (-27)$
46. $(+3) + (+2) + (-5) + (-8)$
47. (a) Add the sum of $+3$ and -5 to -7.
     (b) Add $+3$ to the sum of -5 and -7.
     (c) What law of addition is illustrated by the results of (a) and (b)?

Find the following products:

48. $(+5)(+8)$ **49.** $(+5)(-8)$ **50.** $(-5)(-8)$
51. $(-5)(+8)$ **52.** $(+3)(-6)$ **53.** $(-6)(+6)$
54. $(-7)(+3)$ **55.** $(-7)(-3)$ **56.** $(+16)(-4)$
57. $(-16)(-3)$ **58.** $(-6)(-3)(-2)$ **59.** $(-6)(-3)(+2)$
60. $(-6)(+2)(+2)$ **61.** $(+6)(-4)(-2)(-2)$

Find the following powers:

62. $(+2)^2$ **63.** $(+2)^3$ **64.** $(+2)^4$ **65.** $(+2)^5$ **66.** $(+2)^6$
67. $(-2)^2$ **68.** $(-2)^3$ **69.** $(-2)^4$ **70.** $(-2)^5$ **71.** $(-2)^6$
72. State a rule for writing the sign of a directed number raised to a power.
73. In the figure, A is the position of a car at 12:00 noon.

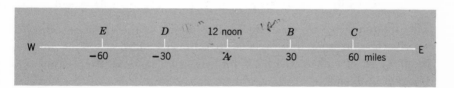

Figure 3

Let: Time before noon be negative.
     Time after noon be positive.
     Position of the car to the right of A be positive.
     Position of the car to the left of A be negative.
     Eastward direction of travel be positive.
     Westward direction of travel be negative.

Using the formula $D = (r)(t)$, that is, distance equals rate times time, find the position of the car at 11 A.M. if it was traveling west at 30 miles per hour.

$$D = (-30)(-1) = 30$$

Thus the car was at point B at 11 A.M., so that moving west at 30 miles per hour, it would reach A at noon.

Using the formula $D = (r)(t)$ and the proper signs for r and t, find the position of the car:

(a) At 3 P.M., if it was traveling west at 20 miles per hour.
(b) At 10 A.M., if it was traveling west at 30 miles per hour.
(c) At 1 P.M., if it was traveling east at 30 miles per hour.
(d) At 9 A.M., if it was traveling east at 20 miles per hour.

6 Subtraction, Division, and Roots of Directed Numbers

Subtraction of Directed Numbers

We have previously described subtraction as the inverse of addition (Section 2). Using this relationship for the subtraction of directed numbers, consider the following examples:

(a) $(+8) - (+3) = +5$ because $(+5) + (+3) = +8$
(b) $(-8) - (-3) = -5$ because $(-5) + (-3) = -8$
(c) $(-8) - (+3) = -11$ because $(-11) + (+3) = -8$
(d) $(+8) - (-3) = +11$ because $(+11) + (-3) = +8$

The difference in each case can be determined as the value necessary to satisfy the corresponding inverse operation of addition. This additive property is, of course, the basis for our usual method of checking subtraction problems.

Let us again consider these four problems with the sign of the number to be subtracted changed from $-$ to $+$, or from $+$ to $-$. If we now add each pair of directed numbers, we have the same results as before. Thus

(a) $(+8) + (-3) = +5$
(b) $(-8) + (+3) = -5$
(c) $(-8) + (-3) = -11$
(d) $(+8) + (+3) = +11$

It is apparent that subtracting a negative number is the same as adding the same positive number; subtracting a positive number is the same as adding the same negative number. We may, indeed, define the subtraction of directed numbers in this manner:

If a and b are any two directed numbers, then by definition

$$a - (-b) = a + (+b)$$
$$a - (+b) = a + (-b)$$

In other words, to subtract one directed number from another, we change the sign of the number to be subtracted, and add. After a little practice, the sign changes should be made mentally. Thus in subtraction

+7	−7	−7	+7	−6	−7	+6	−7	0	0
+3	−3	+3	−3	−6	+7	+8	0	−5	+5
+4	−4	−10	+10	0	−14	−2	−7	+5	−5

Division of Directed Numbers

Since division is the inverse of multiplication, the rule of signs for dividing two directed numbers is the same as the rule of signs for multiplication:

The quotient of two directed numbers having opposite signs is negative and the quotient of two directed numbers having like signs is positive.

Thus

$$(+6) \div (+2) = +3, \quad \text{since} \quad (+2)(+3) = +6$$
$$(-6) \div (-2) = +3, \quad \text{since} \quad (-2)(+3) = -6$$
$$(-6) \div (+2) = -3, \quad \text{since} \quad (+2)(-3) = -6$$
$$(+6) \div (-2) = -3, \quad \text{since} \quad (-2)(-3) = +6$$

Division by Zero

There is one peculiarity of the number zero which is of great importance; *division by zero is never permitted.* The following considerations will show why this is so.

By extending the number system to include positive and negative integers and zero, the sum, difference, and product of any two numbers

a and b is a unique number c.   Stated in another way, there is always one and only one answer when two numbers are added, subtracted, or multiplied.   In the same way, we wish the division of two numbers to give one and only one answer; that is, the result of division is to be unique.   In the quotient $8 \div 2$, the result is 4, which is a unique number. Suppose now that the divisor is zero.   If $8 \div 0 = c$, the definition of division requires that $8 = 0(c)$.   Clearly there is no number c such that $0(c) = 8$.   On the other hand, if both terms of the quotient are zero, so that $0 \div 0 = c$, then we must have $0 = 0(c)$.   Clearly any number c will satisfy this relation, so that while a quotient may be said to exist, it is not unique.   Therefore division by zero is excluded from the operations of mathematics.   This fact must be borne in mind when working with fractions (as in Section 29) and elsewhere where substitution of certain numbers for letters may result in an expression whose denominator equals zero.

Roots of Directed Numbers

What is the value of $\sqrt{+4}$?   Since $(+2)^2 = +4$, we may say that the square root of $+4$ is $+2$.   But it is also true that $(-2)^2 = +4$, so that the square root of $+4$ is also -2.   Thus $+4$ has two square roots; the positive square root we designate by $\sqrt{+4} = +2$, and the negative square root we designate by $-\sqrt{+4} = -2$.

What is $\sqrt[3]{-8}$?   Since $(-2)^3 = -8$, then $\sqrt[3]{-8} = -2$; there is no other value in the set of positive and negative numbers.   Similarly, $\sqrt[3]{8}$ has but one value: $+2$.   Let us now consider $\sqrt{-4}$.   Both $(+2)^2$ and $(-2)^2 = +4$.   Evidently there is no directed number b, positive or negative, such that $\sqrt{-4} = b$.   At present, we must content ourselves with the statement that an even root of a negative number does not exist in the set of directed numbers.

In summary, in the set of directed numbers:

1. A positive number has two square roots, equal in absolute value but opposite in sign.
2. A negative number has no square root within this set.
3. A positive number has one cube root which itself is positive.
4. A negative number has one cube root which itself is negative.
5. Statements 1 and 2 apply to any even root.
6. Statements 3 and 4 apply to any odd root.

EXAMPLE 1. $+\sqrt{729} = 27$

EXAMPLE 2. $-\sqrt{729} = -27$

EXAMPLE 3. $\sqrt[3]{343} = 7$

EXAMPLE 4. $\sqrt[3]{-343} = -7$

EXAMPLE 5. $-\sqrt[3]{-343} = -(-7) = 7$

EXAMPLE 6. $-\sqrt[3]{343} = -7$

Exercise 6

SUBTRACTION, DIVISION, AND ROOTS OF DIRECTED NUMBERS

Find the missing number in each of the following problems:

1. $+7 + (?) = 13$	**2.** $-7 + (?) = 12$	**3.** $-7 + (?) = -12$
4. $-7 + (?) = 0$	**5.** $-3 + (?) = -1$	**6.** $-5 + (?) = -5$
7. $-5 + (?) = +5$	**8.** $-6 + (?) = -3$	**9.** $-8 + (?) = +2$

In each of the following problems, subtract as indicated by considering what number must be added to the first number to get the second number:

10. 4 from 7	**11.** 6 from 17	**12.** 6 from 3
13. 4 from 1	**14.** 4 from -7	**15.** 6 from -17
16. 6 from -3	**17.** 4 from -1	**18.** -4 from 9
19. -6 from 12	**20.** -8 from 3	**21.** -4 from 1
22. -4 from -7	**23.** -6 from -12	**24.** -8 from -3
25. -4 from -1		

In each of the following problems, subtract as indicated by changing (mentally) the sign of the subtrahend and adding the numbers:

26. $6 - (+3)$	**27.** $6 - (-3)$	**28.** $7 - (+18)$
29. $7 - (-18)$	**30.** $12 - (+11)$	**31.** $12 - (-11)$
32. $13 - (+21)$	**33.** $13 - (-21)$	**34.** $-3 - (+5)$
35. $-3 - (-5)$	**36.** $-17 - (+3)$	**37.** $-17 - (-3)$
38. $-25 - (+13)$	**39.** $-25 - (-13)$	**40.** $-76 - (+53)$
41. $-76 - (-53)$		

Find each of the following quotients. Check each answer by multiplication.

42. $\dfrac{+6}{+2}$ 43. $\dfrac{+6}{-2}$ 44. $\dfrac{-6}{-2}$ 45. $\dfrac{-6}{+2}$ 46. $\dfrac{-34}{+17}$ 47. $\dfrac{-34}{-17}$

48. $\dfrac{-30}{-5}$ 49. $\dfrac{+30}{-3}$ 50. $\dfrac{+30}{+3}$ 51. $\dfrac{-40}{-8}$ 52. $\dfrac{-64}{+16}$ 53. $\dfrac{-64}{-16}$

54. $\dfrac{+27}{+9}$ 55. $\dfrac{-27}{+9}$ 56. $\dfrac{-27}{-9}$ 57. $\dfrac{+27}{-9}$ 58. $\dfrac{-35}{-7}$ 59. $\dfrac{-35}{-35}$

Find, by trial, the indicated root of each of the following numbers:

60. $\sqrt{4}$ 61. $-\sqrt{4}$ 62. $-\sqrt{16}$ 63. $\sqrt{16}$ 64. $\sqrt{64}$

65. $\sqrt[3]{8}$ 66. $\sqrt[3]{-8}$ 67. $-\sqrt[3]{27}$ 68. $\sqrt[3]{-27}$ 69. $\sqrt[3]{-64}$

70. $\sqrt[4]{16}$ 71. $-\sqrt[4]{16}$ 72. $-\sqrt[4]{81}$ 73. $\sqrt[4]{81}$ 74. $\sqrt[4]{256}$

75. $\sqrt[5]{32}$ 76. $\sqrt[5]{-32}$ 77. $-\sqrt[5]{243}$ 78. $\sqrt[5]{243}$ 79. $-\sqrt[6]{64}$

Find the value of $a + b$, $a - b$, ab, and $a \div b$ for the following values of a and b, respectively:

80. $-12, -3$ 81. $+6, +2$ 82. $+6, -2$ 83. $-6, -2$

84. $-6, +2$ 85. $+16, +8$ 86. $-16, -8$ 87. $-16, +8$

88. $+6, 0$ 89. $-6, 0$ 90. $0, +4$ 91. $0, -4$

92. $-8, 0$ 93. $0, -6$ 94. $0, 0$

Which of the following operations with zero have no meaning?

95. $\dfrac{6}{0}$ 96. $\dfrac{0}{0}$ 97. $\dfrac{0}{8}$ 98. $\dfrac{0}{2}$ 99. $(6)(0)$

100. $4 + 0$ 101. $0 + 4$ 102. $\dfrac{3}{0}$ 103. $\dfrac{0}{12}$ 104. $0 + 7$

105. $7 - 0$ 106. $0 + 0$

CHAPTER **3**

Algebraic Expressions

7 The Concept of a Variable

Literal Notation

In algebra it is frequently desirable to refer to some numerical quantity without identifying its specific value. At times this may be a matter of convenience, but when we are dealing with a quantity whose value is either unknown or changing it becomes a necessity. In such cases letters are introduced to represent numerical quantities. Thus h may represent the height of a column of mercury in a barometer, N may represent the number of bacteria in a given culture, or x may represent some measurable quantity in a problem situation. We will refer to such letter symbols used to represent numbers as *literal numbers.*

Subscripts or primes are sometimes used with literal numbers to indicate association between them. Thus b and b' (read b and b prime) may be used to represent the bases of a trapezoid, or r_1 and r_2 (read r sub-one and r sub-two) may be used to represent the radii of two concentric circles. In working with such symbols, we must treat them as separate literal quantities; the primes and subscripts are merely a convenience of notation.

The introduction of a literal number into a mathematical discussion makes possible immense savings of mental effort. It enables us to build other expressions by making use of the properties of the given literal number and to perform mathematical operations on the expressions. The purposeful development of such operations and expressions will lead us to two very important applications of algebra—the forming of generalizations and the determination of values which satisfy prescribed conditions in a problem situation.

Constants and Variables

It is instructive to note the role of the literal quantity in each of the following situations. In the statement $n + 4 = 7$, the literal quantity n can assume only one value, namely, $n = 3$. However, in the statement $n + 4 = 4 + n$ (which follows from the commutative property of addition), the value of n can be zero or any positive or negative number. A symbol which represents any unspecified number of a set of numbers is called a *variable*, and the set of numbers is called the *domain* of the variable. When the domain consists of a single element, the symbol is called a *constant*. Thus, in the statement $n + 4 = 7$, n is a constant, and in the statement $n + 4 = 4 + n$, n is a variable. Furthermore, the value of any specific number such as 5, -3, π, or $\frac{1}{2}$ is a constant, whereas the literal quantities in such expressions as $C = \pi d$ or $2x + y - z = 3$ are variables.

At this point it is appropriate to caution against a possible misunderstanding concerning variables. A variable is *not* a quantity which keeps shifting or changing in magnitude as you work with it. When x appears in $(x + 1)^2 = x^2 + 2x + 1$, it takes the place of any number, but it assumes only one value at a time. Thus if x is 8, we have $(8 + 1)^2 = 8^2 + 2 \cdot 8 + 1$, and if x is -3, we have $(-3 + 1)^2 = (-3)^2 + 2(-3) + 1$. Similarly, the statement $y^2 + 14 = 9y$ is true for $y = 7$ or $y = 2$, but y cannot assume a value of 7 in one part of the statement and assume a value of 2 in the remaining part of the statement.

Exercise 7

The Concept of a Literal Number

The following is a list of common formulas. Identify as many as you can by indicating what each variable represents as illustrated in Problem 1:

1. $A = bh$

    Solution: A represents the area of a rectangle or parallelogram.

              b represents the length of its base.

              h represents its height.

2. $A = \frac{1}{2}bh$ **3.** $C = 2\pi r$ **4.** $A = \frac{1}{2}h(b_1 + b_2)$

5. $A = \pi r^2$ **6.** $V = \frac{4}{3}\pi r^3$ **7.** $i = prt$

8. $d = rt$ **9.** $V = lwh$ **10.** $S = 4\pi r^2$

11. $V = \pi r^2 h$ **12.** $V = \frac{1}{3}Bh$ **13.** $a = \dfrac{s}{n}$

14. $d = 2r$ **15.** $p = 4s$ **16.** $c^2 = a^2 + b^2$

17. $p = 2l + 2w$ **18.** $E = IR$ **19.** $s = 16t^2$

State the domain of each variable under the given definition:

20. x represents the number of degrees in one angle of a triangle.
   Solution: $0 < x < 180$.

21. n is the number of days in a calendar year.

22. e represents the even integers between $\frac{1}{2}$ and $9\frac{1}{2}$.

23. l equals the length of the third side of a triangle having two sides equal to 3 inches and 7 inches, respectively.

24. y is the value of the quantity $3x - 2$ as x changes from 0 to 3.

25. z represents the age of a minor child.

26. d is the distance between two distinct points on the circumference of a circle whose radius is 4 feet.

27. M equals the posted maximum speeds for automobiles in your city or state.

28. u represents the length of a straight line connecting any two points on a 3 inch square.

29. v is the numerical value of the expression $x^2 + 6$ as x assumes values between -5 and 5.

Write three statements in each case which result from replacing the literal number by a specific number:

30. $ab = ba$ **31.** $v + v + v = 3v$ **32.** $\dfrac{kn}{n} = k$

33. $7x + x = 8x$ **34.** $\sqrt{a^2b^2} = |ab|$ **35.** $p^2q^2 = (pq)^2$

By systematic trial of numerical values, determine whether the following statements are true for just one or more than one value of the literal number and, therefore, whether the literal numbers represent constants or variables:

36. $3x + 4 = 7x$ **37.** $3x + 4x = 7x$ **38.** $a + 2 = 7$

39. $4r + r = 5r$ **40.** $y = 3y$ **41.** $y = 3y - 2y$

8 Types of Algebraic Expressions

Algebraic Expressions and Terms

The fundamental operations together with the processes of raising to powers and extracting roots are known as *algebraic operations*. When a finite number of algebraic operations are applied to one or more elements of a finite set of explicit numbers and variables, the result is called an *algebraic expression*. To illustrate, consider the set

$$\{+3, -2, x, y, z\}$$

We can combine these elements, through the use of algebraic operations, in a variety of ways. Three examples are

$$3x^2 - 2y + z; \qquad x + \sqrt{3x - y}; \qquad \frac{x + 3y}{x - 2z}$$

Each of these three results is an algebraic expression. If an algebraic expression is regarded as a succession of partial expressions separated by plus and minus signs, each partial expression together with the preceding plus or minus sign is called a *term*. The terms of the expression $3x^2 - 2y + z$ are $3x^2$, $-2y$, and z; the terms of the expression $x + \sqrt{3x - y}$ are x and $\sqrt{3x - y}$; the expression $(x + 3y)/(x - 2z)$ is said to consist of one term.

Terms which have identical literal parts are of special significance since they can be combined. Such terms are called *like terms* or *similar terms*. Thus $3x^2y$ and $-5x^2y$ are like terms, but $3x^2y$ and $-5xy^2$ are not like terms.

Polynomials

A set of algebraic expressions known as polynomials is of particular importance in algebra. A polynomial in x is an algebraic expression whose terms are constants or involve, as variables, only positive integral powers of x. For example, $5x^2$ is a polynomial in x, as is also $4x^3 - 7x^2 - x + 2$. It is apparent that the definition of a polynomial is more restrictive than that of an algebraic expression. For example, the terms $5\sqrt{x}$ and $3/x$ are algebraic expressions but cannot represent terms in a polynomial.

A polynomial in x and y is an algebraic expression whose terms are constants or involve, as variables, only positive integral powers of x and y. Thus $3x^2 + 2xy - 5y^2 + 7x - 2y + 3$ is a polynomial in x and y. Similarly, $x^3 - 2x^2z + yz^2 + x^2 - y + 2z$ is a polynomial in x, y, and z.

If a polynomial consists of a single term, as $-3xy$, it is called a *monomial*. An alternate definition is both evident and useful: a monomial is a term consisting only of the product of positive integral powers of variables and nonzero constants. Any factor of this product is called the *coefficient* of the remaining part. Thus in the monomial $-3xy$, -3 is the coefficient of xy, and $-3y$ is the coefficient of x. The product of the nonzero constants is also called the *numerical coefficient* of the monomial; in common practice it is often referred to simply as the coefficient of the term. Thus -3 is the coefficient of the monomial $-3xy$. If no numerical coefficient appears, as in the term x^2, the coefficient 1 is understood. If no sign is written before a term, as in $4\pi r^2$, the sign $+$ is understood.

Polynomials which contain exactly two terms, such as $3x + 5$, are called *binomials*, and those with exactly three terms, such as $2m^2 - 3mn + 7n^2$, are called *trinomials*. Beyond this point it is not customary to assign special names to polynomials.

Degree of a Polynomial

Since the like terms in a polynomial can be combined, a polynomial is generally expressed as a finite sum of unlike terms. Each term consists of a numerical coefficient multiplied by the product of a finite number of variables and each variable carries an exponent which is a positive integer. The sum of the exponents of the stated variables in any one term is defined as the *degree* of that term in the stated variables. Thus the term $7x^2y$ is of degree three in the variables x and y; it is of degree two in the variable x and it is of first degree in the variable y. The term $-4x^3$ is simply of degree three in x.

The *degree of a polynomial* is the degree of the term of highest degree in the stated variables. Thus the polynomial $7x^2y - 2xy^3 + 4x - 5y$ is of degree four in x and y. It is also of degree two in x and of degree three in y.

It is customary to arrange the terms of a polynomial in order of descending degree in the variables involved. For example, the polynomial

$$4x^3 - 5x^2 + 7x + 8$$

is of degree three in x, and the terms are arranged in descending order. The degree of the term $+8$ is regarded as zero since, as will be shown in Section 44, it can be written in the form $+8x^0$.

Exercise 8

TYPES OF ALGEBRAIC EXPRESSIONS

Identify each of the following as a monomial, binomial, or trinomial:

1. $ax + b$
2. $v - 1$
3. $m_1 x + m_2 y$
4. $y^2 - 2y + 13$
5. $-1v$
6. $b + b'$
7. $\frac{1}{4}\pi d^2$
8. $w^3 + 3w$
9. $3p - 3q - 3$
10. $-\frac{6}{5}y$
11. ax^n
12. $ax^2 + bxy + cy^2$
13. -12
14. $-13r^2 st$
15. $(-2)(4)x - y$

What is the second term in each of the following expressions?

16. $3x^2 - x + 5$
17. $a^2 v^2 - 2av + 7$
18. $P + Prt$
19. $y^3 + 7y^2 - y + 1$
20. $4x^3 + 1$
21. $x^2 + \dfrac{b}{ax} + \dfrac{b^2}{4a^2}$
22. $2 - a$
23. $2\pi r^2 + 2\pi rh$
24. $IR_1 + IR_2 + IR_3$

For each of the following terms list (a) the numerical coefficient and (b) the coefficient of x.

25. $3xy$
26. $\frac{1}{4}xz$
27. $-\frac{5}{2}kx$
28. $-15xy^2$
29. x
30. $\dfrac{x}{4}$
31. $\frac{1}{4}x$
32. $-x$
33. $\dfrac{xy}{z}$
34. $-xy$
35. $2xyz$
36. $\dfrac{-2x}{3}$

Arrange each polynomial in descending powers of the variable and indicate its degree:

37. $x^2 - 5 + x$
38. $m^3 - 8m^2 + m^4 + 10$
39. $4 - 16t^2$
40. $y - y^2$
41. $x^n - 3 + x^{n-1} - 5x^{n-3}$
42. $1 - x - x^2$
43. $3v^2 - 7v + 8 + 4v^3$
44. $3y^{n-1} + y^{n+1} - y$
45. $1 + a^2 + a^4 + a^6$
46. $w + 3 + \frac{1}{4}w^3 + \frac{1}{2}w^2$

What is the degree of each polynomial (*a*) in *x*, (*b*) in *y*, (*c*) in *x* and *y*?

47. $2x^2 - 3xy + y^2$ **48.** $3x^3 - 2x^2y + 7xy^2 - y^3$
49. $4x^2y - 3x + 2y - 5$ **50.** $5x^4 - 4xy^3$
51. $y - y^2 + xy^3$ **52.** xy

Which of the following expressions are not polynomials in *x*?

53. $x^3 - 3x^2 + 6x - 5$ **54.** $3 + \dfrac{3}{x}$ **55.** $7 + 6x - x^2 + \sqrt{x}$

56. $\dfrac{x^3}{4} - \dfrac{3x}{4}$ **57.** $\sqrt{3} - 3x$ **58.** $\sqrt{2x} + 2x$

59. $3 - 3x$ **60.** $\frac{1}{3} - 3x^2$ **61.** $\dfrac{x}{x-1} + \dfrac{x^2}{x+1}$

Indicate the terms which are similar to the first term listed in each line:

62. $5ab$: $5ab^2$, $5a^2b$, $-5ab$, ab, $-2a^2b^2$, $3a$, $5b$, $2abc$, $3ba$, $-ab$, $-17ab$.
63. $-3x^2y$: xy, $-3xy^2$, $3xy^2$, $3xy$, $-3x^2y^2$, $7x^2y$, $\sqrt{2}\,x^2y$, $\sqrt{2x^2y}$, $-3y^2x$.
64. $2uv^2y$: $3uv^2y$, $-uv^2y$, $-2.7uv^2y$, $2u^2vy$, $5uvy^2$, $-2v^2uy$, $4uv^2$, $-u^2vy^2$.

9   The Set of Rational Numbers

Thus far we have included only the positive and negative integers and zero in our number system. For the purpose of evaluating algebraic expressions we will find it useful to extend this number system to include the common fractions and their decimal equivalents. These numbers, the integers and the fractions, both positive and negative, and including zero, comprise the set of *rational numbers*.

We define a rational number as one that can be expressed as the quotient of two integers, that is, a number which can be written in the form *a/b* or $\dfrac{a}{b}$, where *a* is any integer, and *b* is any integer different from zero. Thus each of the numbers

$$\frac{2}{7}, \quad \frac{-5}{6}, \quad \frac{-4}{-9}, \quad 2.07 = \frac{207}{100}, \quad 0.333\cdots = \frac{1}{3}$$

is rational. So also is any integer *n* since it can be written as *n/1*. Therefore the set of rationals contains the system of integers as a subset.

Equality of Rational Numbers

Each rational number can be expressed in many different ways as the quotient of two integers.   For example,

$$\frac{3}{8}, \quad \frac{-6}{-16}, \quad \frac{9}{24}, \quad \frac{21}{56}$$

and different numerals for the same rational number.   Given two symbols for a rational number, we can determine whether they represent the same number by applying the following important definition:

Two rational numbers $\frac{a}{b}$ and $\frac{c}{d}$ are equal if and only if $ad = bc$.

EXAMPLE 1. $\frac{3}{8} = \frac{6}{16}$ because $3 \cdot 16 = 6 \cdot 8$, or 48, and $\frac{11}{16} \neq \frac{12}{17}$ because $187 \neq 192$.

The definition of equality of two rational numbers gives rise to a very useful property,

$$\frac{a}{b} = \frac{ka}{kb} \quad \text{for any integer } k, \ k \neq 0$$

since $a(kb) = b(ka)$.   This property justifies the common procedures of reduction and expansion of fractions, as illustrated in the following examples.

EXAMPLE 2. $\dfrac{34}{51} = \dfrac{17 \cdot 2}{17 \cdot 3} = \dfrac{2}{3}.$

EXAMPLE 3. $\dfrac{5}{7} = \dfrac{4 \cdot 5}{4 \cdot 7} = \dfrac{20}{28}.$

A nonzero rational number p/q is said to be expressed in lowest terms or *simplest form* if the integers p and q have no common factor.   Thus $\frac{15}{18}$ is not in simplest form since both integers have a common factor 3, but $\frac{5}{6}$ is in simplest form since the integers do not have a common factor.

The negative of a rational number a/b is usually expressed as $\dfrac{-a}{b}$, although the forms $\dfrac{a}{-b}$ or $-\dfrac{a}{b}$ are equivalent.   Thus the negative of $\dfrac{5}{6}$ may be written as $\dfrac{-5}{6}, \dfrac{5}{-6},$ or $-\dfrac{5}{6}.$

Operations with Rational Numbers

Addition and subtraction of rational numbers may now be defined by the following relation:

$$\frac{a}{b} + \frac{c}{d} = \frac{ad + bc}{bd}$$

For example,

$$\frac{2}{3} + \frac{1}{7} = \frac{14 + 3}{21} = \frac{17}{21}$$

and

$$\frac{2}{3} - \frac{1}{7} = \frac{2}{3} + \frac{-1}{7} = \frac{14 + (-3)}{21} = \frac{11}{21}$$

Multiplication of rational numbers is defined by the relation

$$\frac{a}{b} \cdot \frac{c}{d} = \frac{ac}{bd}$$

It is apparent that the product of two rational numbers is always a rational number and, interestingly enough, this enables us to develop a necessary definition for the division of rational numbers.   Assume that

$$\frac{a}{b} \div \frac{c}{d} = \frac{p}{q}$$

Then

$$\frac{a}{b} = \frac{c}{d} \cdot \frac{p}{q} \quad \text{or} \quad \frac{a}{b} = \frac{cp}{dq} \quad \text{or} \quad adq = bcp$$

But

$$adq = bcp \quad \text{if} \quad \frac{p}{q} = \frac{ad}{bc}$$

Therefore

$$\frac{a}{b} \div \frac{c}{d} = \frac{ad}{bc}$$

This is consistent, of course, with the common procedure of dividing by a fraction by "inverting the divisor and multiplying."

EXAMPLE 4. $\dfrac{3}{5} \cdot \dfrac{-2}{9} = \dfrac{-6}{45} = \dfrac{-2}{15} = -\dfrac{2}{15}.$

EXAMPLE 5. $\dfrac{2}{7} \div \dfrac{5}{8} = \dfrac{16}{35}.$

Postulates for Rational Numbers

Thus we note that in the system of rational numbers all four fundamental operations are always possible. (Needless to say, division by zero is at all times excluded.) Therefore the set of rationals satisfies the postulates of closure which we introduced in discussing the properties of natural numbers. Similarly, the rational numbers satisfy the commutative, associative, and distributive postulates, as well as the postulates which assert the existence of a zero element and additive inverses. We now list two additional postulates for the set of rational numbers, which we number 10 and 11 to form a continuation of the nine postulates mentioned previously.

Postulate 10. *For any element a in the set, there exists an element E such that $a \times E = E \times a = a$. E is called the multiplicative identity element of the set.*

Postulate 11. *For any element a in the set, $a \neq 0$, there exists an element a' such that $a \times a' = a' \times a = E$. The element a' is called the multiplicative inverse of the element a.*

The multiplicative identity for the set of rational numbers is recognized, of course, as the number 1, and the multiplicative inverse of an element is usually called its reciprocal. Thus the reciprocal of $\frac{5}{7}$ is $\frac{7}{5}$ because $\frac{5}{7} \times \frac{7}{5} = 1$.

If all eleven of these postulates are satisfied by a set of elements, then the set is said to form a *field*. The natural numbers do not form a field, nor do the integers; the set of rational numbers, however, do form a field.

Exercise 9

The Set of Rational Numbers

Write the multiplicative inverse of each of the following rational numbers:

1. 4	**2.** $\frac{2}{5}$	**3.** -6	**4.** $\frac{5}{2}$
5. x	**6.** m/n	**7.** $-p/q$	**8.** $-y$

Write the additive inverse of each of the following rational numbers:

9. 7	**10.** $\frac{3}{4}$	**11.** -9	**12.** $-\frac{1}{5}$

13. Is every integer a rational number? Explain your answer.

14. Is every rational number an integer? Explain your answer.

Express, in simplest form, the rational number which represents each of the following expressions:

15. $\dfrac{2}{3} + \dfrac{1}{2}$

16. $\dfrac{3}{4} - \dfrac{2}{3}$

17. $\dfrac{2}{5} \cdot \dfrac{7}{3}$

18. $\dfrac{4}{9} \div \dfrac{1}{5}$

19. $\dfrac{-5}{6} \cdot \dfrac{2}{5}$

20. $\dfrac{5}{3} \div \dfrac{-1}{3}$

21. $\dfrac{2}{11} + \dfrac{3}{2}$

22. $\dfrac{5}{4} - \dfrac{2}{9}$

23. $\dfrac{12}{5} \cdot \left(-\dfrac{1}{6}\right)$

24. $\dfrac{3}{4} - \dfrac{7}{8}$

25. $\dfrac{1}{4} + \dfrac{1}{5}$

26. $5 \div \dfrac{1}{5}$

27. Which of the eleven field postulates are not satisfied by the set of natural numbers?

28. Which of the eleven field postulates are not satisfied by the set of integers?

29. (a) Can you find the smallest integer that is greater than $\frac{5}{3}$?

(b) Can you find the smallest rational number that is greater than $\frac{5}{3}$? Explain your answer.

30. (a) Can you find the largest integer that is less than $\frac{9}{4}$?

(b) Can you find the largest rational number that is less than $\frac{9}{4}$? Explain your answer.

10 Evaluation of Algebraic Expressions

Substitution

When numerical values of literal quantities are known or assigned, the value of the algebraic expression can be determined. The procedure is one of replacing the literal number by its value and performing the indicated operations. For example, the value of the binomial $x + 7$ when x equals 5 is $5 + 7$ or 12. The process of substitution and evaluation, although very simple in this case, has extensive application in mathematics. It is used to evaluate formulas, check applied problems and equations, find corresponding values of related variables, make estimates, and determine roots of certain equations. It is therefore important to develop ability to work with assurance and accuracy in evaluating algebraic expressions.

Symbols of Grouping

In working with the postulates of addition and multiplication, we noted that parentheses were used to group certain quantities. For example, the associative law of addition

$$a + (b + c) = (a + b) + c$$

indicated that the numbers b and c were to be treated as one group or quantity, and the numbers a and b were to be treated as another group or single quantity. The parentheses in this case represent a symbol of grouping.

Other symbols used to indicate groupings are brackets, [ ], braces, { }, and the vinculum (a bar placed above the grouped quantities). Thus, $a - (b - c)$, $a - [b - c]$, $a - \{b - c\}$, $a - \overline{b - c}$ are equivalent expressions; each indicates that the difference of b and c is to be subtracted from a. The division bar and the radical sign may also serve to indicate groupings, as in the expressions $\sqrt{a + b}$ and $\dfrac{a + b}{2}$.

Symbols of groupings may be used to indicate explicitly the order of fundamental operations in certain expressions. Thus the expression $3 + (5 \cdot 9)$ means that the quantity $5 \cdot 9$ is to be added to 3. The result, of course, is 48. On the other hand, the expression $(3 + 5) \cdot 9$ means that the sum of 3 and 5 is to be multiplied by 9. The result here is 72.

It is not uncommon to encounter several signs of grouping in a single expression. Thus $3[x - 2(x + 6)]$ means that the entire quantity $x - 2(x + 6)$ is to be multiplied by 3, and that within it, the binomial $x + 6$ is to be considered as a single quantity to be multiplied by -2. When two or more signs of grouping are used, one within the other, it is best to simplify them one at a time beginning with the innermost. For example, if $x = 7$, the expression

$$3[x - 2(x + 6)]$$

becomes

$$3[7 - 2(7 + 6)]$$

which simplifies in successive steps

$$3[7 - 2(13)]$$
$$3[7 - 26]$$
$$3[-19] = -57$$

Exercise 10

EVALUATION OF ALGEBRAIC EXPRESSIONS

Find the value of each of the following algebraic expressions if $x = 10$ and $y = 2$:

1. $x - y$
2. $\dfrac{x}{y}$
3. $5xy$

4. $2x - 3y$
5. $7x - 5xy$
6. $2x + 4xy - y$

7. $\dfrac{9y}{x}$
8. $\dfrac{3x + 6y}{7}$
9. $\dfrac{2xy + 3y - 5}{3xy - 2x - 3}$

Evaluate each expression for $y = -4$:

10. $3(y - 2)$
11. $2(3y + 5)$
12. $y(4 - y)$
13. $3y(2y - 1)$
14. $-5y(6 + 2y)$
15. $5 + 3(y + 7)$
16. $y + 2 - (3y + 9)$
17. $y + 6(y - 2)$
18. $(y + 6)(y - 2)$

What is the value of each of the following expressions for $x = 3$, $y = 5$?

19. $2[x - (2y + 3)]$
20. $3x^2 - [3 - (2x - 1)]$
21. $x - [x - (x^2 - y)]$
22. $x^2 - [2x - (3x - \overline{y + 1})]$
23. $(x^2 - 2)(x - 1)$
24. $4 - [x + 2(x - y)]$
25. $(x^2 + x - 1)(2x + 3)$
26. $(y^2 - y)(3y + 2)$
27. $(x - y)^3$
28. $\sqrt{x^2} + \sqrt{y^2}$

Show that the following expressions are equal for one of the listed values and are unequal for the remaining values of the variable:

29. $3(x + 7) - 2$ and $6x + 10$ for $x = 5, 3, 1$.
30. $3x^2 + 7x + 5$ and 11 for $x = 3, -3, -\frac{2}{3}$.
31. $y^4 - y^3 + 2y^2 - 4y - 8$ and 0 for $y = 1, -1, -2$.
32. $2r^3 + 3r^2$ and $4r - 1$ for $r = 1, -\frac{1}{2}, \frac{1}{2}$.
33. $\dfrac{b + 10}{b^2 - b}$ and $\dfrac{b - 1}{b} - \dfrac{b}{b - 1}$ for $b = 2, 3, -3$.
34. $v - 2(2 - 3v)$ and $3(v + 4)$ for $v = 0, \frac{1}{2}, 4$.
35. $\frac{1}{4}(2z - 5)$ and $\frac{1}{2}z + \frac{5}{6}(z + 1)$ for $z = 1, \frac{5}{2}, -\frac{5}{2}$.
36. $\dfrac{3}{t + 3}$ and $\dfrac{2t}{t - 2} - 2$ for $t = -9, -18, 18$.
37. $(2x - 3)^2$ and $4x^2 + 2(7 - x)$ for $x = -\frac{1}{2}, \frac{1}{2}, \frac{1}{3}$.
38. $x - 2$ and $2\sqrt{x - 2}$ for $x = 6, 11, 18$.

The following expressions are true for all admissible values of the literal number.   Check the equality by substituting any specific number for the literal quantity.

39. $(3x - 1)^2 + 6x = 9x^2 + 1$ **40.** $\dfrac{y^2 + 2y - 3}{y - 1} = y + 3$

41. $\dfrac{t^3 + 1}{t + 1} = t^2 - t + 1$ **42.** $v - 3(1 - v) = 4(v - 1) + 1$

Evaluate each expression for $x = -3, -2, -1, 0, 1, 2, 3$, and determine whether the expression, for this range of values, is increasing in value, decreasing in value, or both:

43. $3x + 5$ **44.** $-4x + 2$ **45.** $2x - 3$

46. $-x + 15$ **47.** $x^2 - x + 5$ **48.** $-3x^2 + 2x - 2$

49. $x^2 + 10$ **50.** $-x^2 + x$ **51.** x^3

Evaluate the following formulas:

52. $E = IR$ for $I = 7.5$, $R = 14$

53. $A = p + prt$ for $p = 250$, $r = 0.05$, $t = 6$

54. $W = \dfrac{wd}{D}$ for $w = 100$, $d = 3$, $D = 5$

55. $s = \frac{1}{2}at^2$ for $a = 32$, $t = 3$

56. $C = \frac{5}{9}(F - 32)$ for $F = 122$

57. $p = 2(l + w)$ for $l = 3.7$, $w = 2.6$

58. $V = \frac{4}{3}\pi r^3$ for $\pi = \frac{22}{7}$, $r = 3$

59. $t = \dfrac{d}{r}$ for $d = 1400$, $r = 40$

60. $v = V + gt$ for $V = 30$, $g = 32$, $t = 5$

61. $s = -16t^2 + v_0 t + h$ for $t = 5$, $v_0 = 1600$, $h = 300$

By using signs of grouping, show that:

62. $3x - 7$ is to be subtracted from $5y$.

63. The sum of x and y is to be multiplied by c.

64. Three times the difference of 8 and x is to be added to $4a$.

65. Four times the sum of $7a$ and $3b$ is to be divided by the sum of a and b.

66. The difference of $6x$ and 4 is to be multiplied by x.

67. Show that the operation of division is not associative, that is, that $(a \div b) \div c$ is not always equal to $a \div (b \div c)$.

11   Operations with Monomials

Addition of Monomials

It has been previously stated that only like quantities can be added. In the case of monomials, this means quantities with identical literal parts, that is, like terms.   For example, $5ax$ and $6ax$ are like terms and their sum may be obtained as follows:

$$5ax + 6ax = ax \cdot 5 + ax \cdot 6 \qquad \text{(commutative law of multiplication)}$$
$$= ax(5 + 6) \qquad \text{(consequence of the distributive law)}$$
$$= (5 + 6)ax \qquad \text{(commutative law of multiplication)}$$
$$= 11ax \qquad \text{(closure law of addition)}$$

Thus we note that to add two like terms, we add the numerical coefficients and assign the common literal factor.

The addition of terms which are not similar can only be indicated; their sum cannot be written as a single term.   For example, the sum of $3x$ and $-4y$ is $3x - 4y$.   To *combine* terms in an expression means to add the similar terms.   Thus combining terms in the polynomial $12u - 3v - 4u + 6$ results in $8u - 3v + 6$.

If a series of similar terms, both positive and negative, are to be combined, it is often expedient to add the positive and negative terms separately and then combine the two partial sums.   For example, $3x + 8x - 2x + 4x - 6x$ becomes $15x - 8x$ or $7x$.   A convenient method for checking the result of a problem in addition is to perform the addition in some other order.

Subtraction of Monomials

Subtraction, like addition, can be performed only with like terms.   To subtract two monomials, we convert the problem to addition (see the definition of subtraction, Section 6) and then proceed as in the preceding section.   For example,

$$v^2 - (-5v^2) = v^2 + (+5v^2) = 6v^2$$

Laws of Exponents

Before continuing with the remaining algebraic operations with monomials it is necessary to introduce several important theorems concerning exponential quantities.   These theorems, known as the *laws of exponents* will be immediately useful to us in our further discussion of operations with algebraic expressions and later in the development of properties of exponents outside the set of natural numbers.

It will be recalled that the exponential expression b^n has been defined to represent the product of n factors each equal to b.   As a consequence of this definition we now state and prove the following theorems.

Theorem 1.   If m and n are positive integers, then

$$b^m \cdot b^n = b^{m+n}$$

Proof. $b^m = b \cdot b \cdot b \cdots b$ (m factors of b)

$b^n = b \cdot b \cdot b \cdots b$ (n factors of b)

$b^m \cdot b^n = (m \text{ factors of } b)\ (n \text{ factors of } b)$

$= \overline{(m + n \text{ factors of } b)}$

$= b^{m+n}$

Theorem 2.   If m and n are positive integers and $b \neq 0$, then

$$\frac{b^m}{b^n} = b^{m-n} \quad (m > n) \qquad \text{or} \qquad \frac{b^m}{b^n} = \frac{1}{b^{n-m}} \quad (m < n)$$

Proof.   For $m > n$,

$$\frac{b^m}{b^n} = \frac{b \cdot b \cdot b \cdots b}{b \cdot b \cdot b \cdots b} \qquad \begin{matrix}(m \text{ factors of } b) \\ (n \text{ factors of } b)\end{matrix}$$

After dividing numerator and denominator by the factor b, n times, we are left with $(m - n)$ factors in the numerator. Hence,

$$\frac{b^m}{b^n} = \frac{b^{m-n}}{1} = b^{m-n}$$

For $m < n$, the proof follows a similar form.

Theorem 3.   If m and n are positive integers, then

$$(b^m)^n = b^{mn}$$

Proof. $(b^m)^n = b^m \cdot b^m \cdot b^m \cdots b^m$ (n factors of b^m)

$= b^{m+m+m+\cdots+m}$ (n terms in the exponent, each equal to m)

$= b^{mn}$

Theorem 4.   If n is a positive integer, then

$$(ab)^n = a^n b^n$$

Proof. $(ab)^n = (ab)(ab)(ab) \cdots (ab)$ (n factors of ab)
$$= (a \cdot a \cdot a \cdots a)(b \cdot b \cdot b \cdots b)$$

     (n factors of a and n factors of b)

$$= a^n b^n$$

EXAMPLE 1. $y^3 \cdot y^4 = y^7$ by Theorem 1

EXAMPLE 2. $(5^3)^4 = 5^{12}$ by Theorem 3

EXAMPLE 3. $\dfrac{x^5}{x^2} = x^3$ by Theorem 2

EXAMPLE 4. $(3c^2)^3 = 27(c^2)^3$ by Theorem 4
$$= 27c^6$$ by Theorem 3

Multiplication of Monomials

Consider the product $(-5x^3y^2)(3x^2y)$.   Since multiplication is commutative we can rearrange the factors and apply Theorem 1 of the laws of exponents as follows:

$$(-5x^3y^2)(3x^2y) = -5 \cdot 3 \cdot x^3 \cdot x^2 \cdot y^2 \cdot y = -15x^5y^3$$

Note that the numerical coefficient of the product is obtained by multiplying the numerical coefficients of the given monomials, and the exponents of the literal factors are determined by applying the appropriate law of exponents.   Thus, in multiplication

$-3ax^2$	$-5y^2$	$7bt^2$	$9a^4c^2d^3$
$4ay$	$-z$	-2	$6ac^3$
$-12a^2x^2y$	$5y^2z$	$-14bt^2$	$54a^5c^5d^3$

Division of Monomials

Since division is the inverse of multiplication, the procedure for dividing monomials can be developed in a similar manner to that used for multiplication.   Thus

$$\frac{18a^2c^3}{-2a^2c} = \frac{18}{-2} \cdot \frac{a^2}{a^2} \cdot \frac{c^3}{c} = -9c^2$$

Therefore the quotient can be obtained by dividing the numerical coefficients and using Theorem 2 of the law of exponents to determine the exponent of like literal factors. If like literal factors have equal exponents, the expression may be simplified by dividing both numerator and denominator by that factor. Thus, in division

$$\frac{14bt^2}{2bt^2} = 7 \qquad \frac{-12x^2y^2z}{-3xz} = 4xy^2 \qquad \frac{5av^3}{a} = 5v^3 \qquad \frac{3p^2q^3}{-3p^2q^3} = -1$$

The usual check for division is to multiply the quotient by the divisor to obtain the dividend.

Powers of Monomials

A power of a monomial can be found by applying the laws of exponents directly. For example:

$$(-5x^2y)^3 = (-5)^3(x^2)^3(y)^3 \qquad \text{by Theorem 4}$$
$$= -125x^6y^3 \qquad \text{by Theorem 3}$$

Roots of Monomials

We have seen previously (Section 6) that a positive integer has two square roots. For example, the square roots of 49 are $+7$ and -7. The radical sign was used to indicate each root as follows:

$$+ \sqrt{49} = +7$$
$$- \sqrt{49} = -7$$

We now wish to extend the symbolism of the radical sign to include the indicated root of certain algebraic expressions.

If b is any algebraic expression, the symbol $\sqrt[n]{b}$ is said to represent the *principal nth root of b*. The principal nth root of b is defined to be positive if b is positive; it is defined to be negative only if b is negative and n is odd. In particular:

1. The principal square root of b is positive if b is positive, and it is not defined if b is negative.

2. The principal cube root of b is positive if b is positive and negative if b is negative.

Therefore for positive values of the literal numbers it follows that

$$\sqrt{64u^6} = 8u^3$$
$$-\sqrt{64u^6} = -8u^3$$
$$\sqrt[3]{64u^6} = 4u^2$$
$$\sqrt[3]{-64u^6} = -4u^2$$

Each of these results can be verified by using the relationship that finding powers and extracting roots are inverse operations. For example,

$$\sqrt{64u^6} = 8u^3 \qquad \text{because} \quad (8u^3)^2 = 64u^6$$

and

$$\sqrt[3]{-64u^6} = -4u^2 \qquad \text{because} \quad (-4u^2)^3 = -64u^6$$

This relationship between operations is the basis for the following procedure for finding an indicated root of a monomial expression.

1. Find the indicated root of each factor.
2. If a factor is an exponential quantity, divide the exponent by the root index.

Therefore $\sqrt[3]{-125x^6y^3}$ is considered as $\sqrt[3]{-125} \cdot \sqrt[3]{x^6} \cdot \sqrt[3]{y^3}$, which simplifies to $-5x^2y$. Similarly, $\sqrt{9c^4y^8} = \sqrt{9} \cdot \sqrt{c^4} \cdot \sqrt{y^8} = 3c^2y^4$ and $\sqrt[3]{27a^{12}b^9} = 3a^4b^3$.

Exercise 11

Operations with Monomials

Write each expression in its most concise form using exponents:

1. $4xxxx$
2. $3xyy$
3. $-12abbc$
4. $-uuuv$
5. $2rtrts$
6. $-7aabbccc$

Write each expression without the use of exponents:

7. $2x^2y^2$
8. $5a^2c^3$
9. $-3p^3q^4$
10. $11a^3bc^2$
11. $-u^2v$
12. $5ax^3$

Find each product by first writing the expressions without using exponents, as in Problem 13:

13. $(2a^2c)(6abc) = 2aac \cdot 6abc = 2 \cdot 6aaabcc = 12a^3bc^2$
14. $(3rs^2)(8r^2st)$
15. $(7xy^2z^3)(-2xy^3z^3)$
16. $(-4m^3n)(-m^2n^3)$

Find the following products directly, as in Problem 17:

17. $(5xy^3)(x^2y) = 5x^3y^4$ 18. $(-8a^2b)(3ab^2)$ 19. $(8mn)(-3mn^2)$
20. $(-x)(-y)$ 21. $(-1)(-b^2cd^3)$ 22. $4(-3abc)$
23. $(r)(r^4)$ 24. $13x \cdot 2x$ 25. $-5y \cdot 3y^2$
26. $-6v^2 \cdot 3v^3$ 27. $-8w(-3w^5)$ 28. $2^3 \cdot 2^4$
29. $5 \cdot 5^5$ 30. $7^2 \cdot 7^4$ 31. $4^4 \cdot 4^4$

32. $\quad x^2$ 33. $-3a^2$ 34. $-4xy^2$
$\quad\underline{-x^3}$ $\underline{-2a^2}$ $\underline{2x^2y}$

35. $\quad -7axy^2$ 36. $\quad 4a^2bcx^3$ 37. $-8n^2x^3$
$\quad\underline{-3a^2y^2}$ $\quad\underline{-3adxy^3}$ $\underline{2mnx}$

38. $5a^2bc^2(-b^3x)$ 39. $(-11dm^2x^3)(-6d^3x^2y)$
40. $r^2 \cdot r \cdot r^4$ 41. $x^2y \cdot x^2y^2 \cdot xy^3$
42. $(-x)(-x)(-x)$ 43. $(-5ac)(-cd)(-6bc)$

Find the following quotients by first writing the expressions without using exponents, then dividing numerator and denominator by like factors, as in Problem 44:

44. $\dfrac{-15u^2vw^3}{-5uw^3} = \dfrac{-15\cancel{u}uv\cancel{w}\cancel{w}\cancel{w}}{-5\cancel{u}\cancel{w}\cancel{w}\cancel{w}} = 3uv$ 45. $\dfrac{32m^4n^5}{4m^4n}$

46. $(18y^2z^5) \div (-3yz^4)$ 47. $(x^7y^6) \div (x^5y^2)$ 48. $(-14a^4b^2) \div (-ab^2)$

Find the following quotients directly, as in Problem 49. Check each problem.

49. $(-9x^3y^5) \div (9xy) = -x^2y^4$ 50. $(-22cd^2) \div (-11cd)$
51. $45b^5m^3t^2 \div 15b^5m^3t^2$ 52. $-20m^3 \div 5m^2$
53. $-18n^2 \div 9n^2$ 54. $-35w^5 \div 7w^3$
55. $(-x^3) \div (-x^2)$ 56. $(-13rlt^2) \div (-13rlt^2)$
57. $(8ab^2x) \div (-8ab^2x)$ 58. $3z \div 3z$
59. $-9ac \div 9ac$ 60. $(-y^9) \div (-y)$
61. $a^7 \div a^6$ 62. $m^4 \div m^4$

Add the following:

63. $\quad -4u$ 64. $\quad -8x$ 65. $\quad 9v$ 66. $\quad -5y$
$\quad\underline{-3u}$ $\quad\underline{7x}$ $\quad\underline{-12v}$ $\quad\underline{2y}$

67. $\quad -2b$ 68. $\quad 2dx$ 69. $-3a^2b^2x$ 70. $\quad 8a^2m$
$\quad\underline{b}$ $\quad\underline{-2dx}$ $\underline{10a^2b^2x}$ $\quad\underline{-6a^2m}$

71. $9x^5y^2z$
$8x^5y^2z$

72. $-5bm^2x$
$-5bm^2x$

73. b
$-b$

74. $-2pq$
$5pq$

75. $3(a + b)$
$5(a + b)$

76. $-9(x - y)$
$-6(x - y)$

77. $6(2c - d)$
$-8(2c - d)$

78. $-12(a^2 + 2ax)$
$4(a^2 + 2ax)$

Subtract and check:

79. $3ab$
$7ab$

80. $2x^2$
$-7x^2$

81. $8y^3$
$-3y^3$

82. $-11at$
$13at$

83. $7(a - 2b)$
$-8(a - 2b)$

84. $14(x^2 + x)$
$4(x^2 + x)$

85. $-2(p + q)$
$-(p + q)$

86. $-3(c - d^2)$
$8(c - d^2)$

87. $3x - (2x)$

88. $3x - (-2x)$

89. $(-4mn) - (-2mn)$

90. $(2pt) - (7pt)$

91. $(t^2) - (t^2)$

92. $(-11cd) - (-cd)$

Add the following monomials:

93. $-x$
$5x$
$-x$

94. $-2xy^2$
$6xy^2$
$-8xy^2$

95. a^2m^2x
a^2m^2x
$-3a^2m^2x$

96. $5uv$
$2uv$
$-6uv$

97. $3b, -b, -2b, 6b, b, 3b, -2b$

98. $-2m^2, 6m^2, -7m^2, -8m^2, 9m^2, m^2$

99. $4ab, -5ab, -3ab, 2ab, -8ab$ **100.** $x, 2x, -8x, 11x, -5x, 3x, 8x$

Find the indicated power of each monomial:

101. $(2b^4)^2$

102. $(-2u^2)^3$

103. $(3v^3)^3$

104. $(-a^3y^2)^2$

105. $(-5x^5)^2$

106. $(5cd^2)^3$

107. $(-6bx^3)^3$

108. $(3a^4x^6)^4$

109. $(-9ab^2c)^3$

110. $(-4ax^2y^6)^3$

111. $(-10x^3yz^4)^3$

112. $(-10cd)^2$

113. $(-a)^2$

114. $(7c)^2$

115. $(-3b^3)^3$

116. $(3p^2q)^4$

Determine the indicated root of each monomial for positive values of the literal numbers and check:

117. $\sqrt{49a^6b^4}$

118. $\sqrt{x^2}$

119. $\sqrt{144b^8x^{14}y^{10}}$

120. $\sqrt{4x^4}$

121. $\sqrt[3]{t^6}$

122. $\sqrt[3]{8b^3d^{12}}$

123. $\sqrt[3]{125a^9x^3}$

124. $\sqrt[3]{-27y^6}$

125. $\sqrt[4]{r^4s^{16}}$

126. $\sqrt[5]{32y^{10}}$

127. $\sqrt{1.69a^4b^2}$

128. $\sqrt[3]{-64z^3}$

12 Operations with Polynomials

Addition of Polynomials

The procedure for adding polynomials is based on the procedure for adding monomials. The commutative and associative laws are used to group like terms. To illustrate, let us add the following polynomials:

$$(3x^2 - 2y^2 + 5x - 7) + (x^2 - 7xy + 8y^2 - 2x + 4)$$

Using the commutative and associative laws to group like terms we have:

$$(3x^2 + x^2) + (-7xy) + (-2y^2 + 8y^2) + (5x - 2x) + (-7 + 4)$$

Now following the procedure for adding monomials we combine like terms to obtain the final result:

$$4x^2 - 7xy + 6y^2 + 3x - 3$$

For convenience we may arrange the like terms in columns. Thus for the example above we write

Check:

Let $x = 2$, $y = 3$. Then

$$\begin{array}{l} 3x^2 \qquad\quad - 2y^2 + 5x - 7 \\ \underline{x^2 - 7xy + 8y^2 - 2x + 4} \\ 4x^2 - 7xy + 6y^2 + 3x - 3 \end{array}$$

$$\begin{array}{rl} 3x^2 \qquad\quad - 2y^2 + 5x - 7 = & -3 \\ \underline{x^2 - 7xy + 8y^2 - 2x + 4 =} & \underline{+34} \\ 4x^2 - 7xy + 6y^2 + 3x - 3 = & +31 \end{array}$$

The form of check illustrated, although long, has some merit in that it provides excellent practice in evaluation of algebraic expressions. A more common method of checking addition is one mentioned previously—merely add the terms in some order different from the order used to obtain the sum initially.

Subtraction of Polynomials

To subtract one polynomial from another we use the definition of subtraction (Section 6) to convert the problem to addition and then proceed as in addition of polynomials. For example,

$$(3a + 7b - 12) - (-5a - 8b)$$

is equal to the sum

$$(3a + 7b - 12) + (5a + 8b) = 8a + 15b - 12$$

Note that the sign of *each* term in the polynomial to be subtracted is changed when the problem is converted to addition.   As with addition, the like terms may be arranged in columns for convenience.

Multiplication of Polynomials

The multiplication of a polynomial and a monomial is a direct application of the distributive law, which, it will be recalled, takes the form $a(x + y) = ax + ay$.   Therefore   the   product $3u(2v - 9uv)$ becomes $(3u)(2v) + (3u)(-9uv)$ and, multiplying the monomials, this simplifies to $6uv - 27u^2v$.   This plan of multiplication may be extended to yield the product of a monomial and a polynomial of any number of ter s. In each example the product will be equal to the sum of all the partial products obtained by multiplying each term of the polynomial by the monomial.   For example,

$$2x(x^2 - 2xy + 5y^2) = 2x^3 - 4x^2y + 10xy^2$$

and

$$-3ab^2(a^3 - 3a^2 + a - 7) = -3a^4b^2 + 9a^3b^2 - 3a^2b^2 + 21ab^2$$

The product of two binomials also follows as a consequence of the distributive law.   If in the statement $a(x + y) = ax + ay$, we let a represent the expression $(u + v)$, then

$$(u + v)(x + y) = (u + v)x + (u + v)y = ux + vx + uy + vy$$

Note that the product of two binomials is the sum of all the partial products obtained by multiplying each term of one binomial by each term of the other binomial.

In like manner, the distributive law can be extended to yield the product of any two polynomials.   For convenience the partial products may be developed and arranged as in the following example.

EXAMPLE 1.   Multiply $x^2 - 4x + 7$ by $3x - 5$.

$$
\begin{array}{l}
x^2 - 4x + 7 \\
 3x - 5 \\
\hline
3x^3 - 12x^2 + 21x \qquad \text{(each term multiplied by } 3x) \\
 - 5x^2 + 20x - 35 \quad \text{(each term multiplied by } -5) \\
\hline
3x^3 - 17x^2 + 41x - 35 \quad \text{(like terms combined)}
\end{array}
$$

Division of Polynomials

The division of polynomials falls into two sections: the division of a polynomial by a monomial, and the division of a polynomial by a polynomial. The procedure for the first is again related to the distributive law. Since division is the inverse of multiplication, the distributive law $a(x + y) = ax + ay$ leads to the relationship

$$(ax + ay) \div a = x + y$$

Generalizing this relationship we observe that to divide any polynomial by a monomial we divide *each* term of the polynomial by the monomial and form the algebraic sum of the quotients obtained. Thus

$$
\begin{aligned}
&(5x^4 - 10x^3 + 5x^2) \div 5x^2 \\
&= (5x^4 \div 5x^2) + (-10x^3 \div 5x^2) + (5x^2 \div 5x^2) \\
&= x^2 - 2x + 1
\end{aligned}
$$

To check the division, note that $5x^2(x^2 - 2x + 1) = 5x^4 - 10x^3 + 5x^2$.

We now consider the division of one polynomial by another. Let us first form the product

$$(x^2 - 4x + 7)(3x - 5) = 3x^2 - 17x^2 + 41x - 35$$

It is evident, from the meaning of division, that the product divided by either factor will result in the remaining factor, for example,

$$(3x^3 - 17x^2 + 41x - 35) \div (3x - 5) = x^2 - 4x + 7$$

Our objective is to obtain this result without resorting to previous knowledge of the multiplication relationship. We illustrate the procedure by an example.

EXAMPLE 2.   Divide $3x^3 - 17x^2 + 41x - 35$ by $3x - 5$.

$$
\begin{array}{r}
x^2 - 4x + 7 \qquad \text{(quotient)} \\
\text{(divisor) } 3x - 5 \overline{)3x^3 - 17x^2 + 41x - 35} \quad \text{(dividend)} \\
\underline{3x^3 - 5x^2} \\
-12x^2 + 41x \\
\underline{-12x^2 + 20x} \\
21x - 35 \\
\underline{21x - 35}
\end{array}
$$

Explanation. 1. Divide the first term of the dividend by the first term of the divisor to obtain the first term of the quotient:

$$3x^3 \div 3x = x^2$$

2. Multiply the entire divisor by the first term in the quotient and write the product under the similar terms of the dividend:

$$x^2(3x - 5) = 3x^3 - 5x^2$$

3. Subtract to obtain the new dividend:

$$-12x^2 + 41x - 35$$

4. Divide the first term of the new dividend by the first term of the divisor to obtain the next term of the quotient:

$$-12x^2 \div 3x = -4x$$

5. Multiply the entire divisor by the second term in the quotient, subtract, and so continue until either the remainder is zero or until the remainder can no longer be divided by the divisor.

It will sometimes be necessary to first arrange the dividend and divisor in the order of descending powers of the literal number. Furthermore, if there are missing powers of the literal number, it may be necessary to provide space for them in the dividend. These two points are illustrated in the following example.

EXAMPLE 3. Divide $25 + 6a^3 - 14a$ by $2a + 4$.

$$
\begin{array}{r}
3a^2 - 6a + 5 \\
2a + 4 \overline{)6a^3 + 0a^2 - 14a + 25} \\
\underline{6a^3 + 12a^2} \\
-12a^2 \\
\underline{-12a^2 - 24a} \\
10a \\
\underline{10a + 20} \\
5 \quad \text{(remainder)}
\end{array}
$$

Since there is a remainder, the divisor $2a + 4$ is not an exact factor of the dividend. The relationship is

$$\frac{6a^3 - 14a + 25}{2a + 4} = 3a^2 - 6a + 5 + \frac{5}{2a + 4}.$$

multiples of 3

Exercise 12

Operations with Polynomials

Add and check:

1. $3x - 4$
$2x + 5$

2. $5a + 2$
$-4a - 1$

3. $3v - 2w$
$2v - 3w$

4. $7y + z$
   $y - z$

5. $6x^2 - 9y^2$
$-6x^2 - 9y^2$

6. $3c^2 - 2cd - 4d^2$
$2c^2 - 7cd + 8d^2$

7. $3a - 2b + 6c$
$7a + 5b - 4c$
$2a + 7b + c$

8. $10u - 3v + 9w$
$-6u - 8v - w$
   $u + v$

9. $4x^2 - 8x + 7$
$2x^2 + 7x - 9$
   $x^2 - 4x + 6$

Find the sum of:

10. $3x^2 - 2x + 8, 4x^2 - 5, 6x + 2$

11. $9a^2 - 5b^2, 2c^2 - 4bc, -6c^2 - bc + c^2, a^2 - b^2 + 7c^2$

12. $7ab - 2ab^2 + 7ab^3, 6ab^2 - ab - 2ab^3, 9ab^3 - 7ab^2$

Simplify:

13. $7x^2 - 3xy - 2y^2 - 5xy - x^2 - 3x^2 - 4xy + y^2$

14. $8b^3 + b^2 + 5 - b - 3b^2 + 3b + 3 - 8b^2 - 7b + 12b^2$

15. $p - 2q - 3r - 4 + 7 + 5q - 6r + 7p - 9q + 6p - 4r - 6$

16. $a^2b - 3ab^2 - 7ab - 1 + 5ab^2 - a^2b - 1 + 3ab - 4 + 2ab + 6ab^2$

17. $7m^2n^2 + 8m^2n - 6mn^2 + 3m^2n - 5mn + 8mn^2 - m^2n^2 - 2mn$

Subtract and check:

18. $2r - 5$
$4r + 2$

19. $5xy - 3x$
$-5xy + 3x$

20. $8p^2 + pq - 3q^2$
$2p^2 - 4pq + 5q^2$

21. $7x + 8$
$7x - 8$

22. $5m^2 + mn - 3n^2$
          mn

23. $6x^3 \qquad\qquad - 1$
$-2x^3 + 7x^2 - 8x + 4$

Subtract as indicated:

24. $(y^2 - y) - (y + 2)$ **25.** $(x + y) - (x - y)$

26. $(p^2 - 3p + 6) - (2p^2 + 7p - 4)$

27. $(2 - x + 3x^2) - (x^2 - 4 - 2x)$

28. From $5a + 2$ subtract $3a - 1$.

29. Subtract $6x^3 - 5x^2 - x$ from $2x^3 + 7x^2 - x$.

30. Find the difference between $3c^2 + 2cd + 8d^2$ and $c^2 - cd + d^2$.

31. From the sum of $7a$ and $2b$, subtract their difference.

32. What must be added to $2a + 3b - 4c$ to produce $-a - 7b - 6c$?

33. What must be added to x to produce $-2y$?

Multiply to remove parentheses and then combine like terms:

34. $\quad 3(\ x^2 - 2xy + \ y^2)$
$\quad\quad -5(\ 3x^2 - 2xy - 2y^2)$
$\quad\quad \underline{-4(-x^2 + \ \ xy - 3y^2)}$

35. $\quad 2a(\ a + \ b - 3)$
$\quad\quad 7a(3a - 2b + 1)$
$\quad\quad \underline{-6a(\ a - 8b - 5)}$

36. $3(7a - 2b + 3c)$
$\quad \underline{1(6a + 6b - \ \ c)}$

37. $7(4x - 5y)$
$\quad \underline{5(5x + 7y)}$

38. $x^2 - 8xy + 7y^2 - 5x + 3y - 2 + 4(3x^2 + y^2 - 8x + 3)$

39. $4x - 12y + 7z - 3(2x - 4y + z)$

Find the following products, and check by substituting numbers for the literal quantities (avoid the values 0 and 1 in making substitutions):

40. $(x + 3)(x + 2)$

41. $(2a - 5)(2a - 5)$

42. $(y + 7)(y - 5)$

43. $(3x - y)^2$

44. $(2p - 3)(p + 5)$

45. $(p + 7q)(4p + 8q)$

46. $(m + 2n)(6m - 2n)$

47. $(7x - 4)(7x + 4)$

48. $(7u - 2v)(u + 3v)$

49. $(6 - 5b)(5b - 1)$

50. $(x^2 + 2x + 1)(x + 1)$

51. $(4a^2 + a + 4)(2a - 3)$

52. $(3y^2 - 4y + 1)^2$

53. $(x^2 + 3x^3 - 5x + 1)(3x^2 - 2)$

54. $(7v^3 - 8v^2 + 4v - 1)(2v^2 - 3v + 5)$

Divide and check by multiplication:

55. $(a^3 - a^2) \div a$

56. $(8x^3 - 16x^2y - 12x^2) \div 4x^2$

57. $(a - b + c) \div (-1)$

58. $(-21y^4 - 35y^3 - 7y^2) \div (-7y^2)$

59. $(x^2y^2 - 9x^4y^3) \div (-xy^2)$

60. $(2\pi r^2 + 2\pi rh) \div 2\pi r$

61. $(12a^2b^3 - 36a^4b^3 - 24a^3b^2 + 18ab^2) \div 6ab^2$

62. $(7a^4y - 14a^3y^2 + 35a^2y^3 - 14ay^4) \div (-7ay)$

63. $(x^3 - x^2 + x) \div (-x)$

64. $(6r^2t - 4rt + 8rt^2) \div (-2rt)$

Using division, determine whether:

65. $(x - 1)$ or $(x - 2)$ are exact factors of $x^3 - 6x^2 + 9x - 2$.

66. $(x - 1)$ or $(x + 1)$ are exact factors of $x^3 - 3x^2 - 3x + 1$.

67. $(x - 1)$, $(x - 2)$, or $(x - 3)$ are exact factors of $x^3 + x^2 + x + 6$.

68. $(x - 1)$, $(x + 1)$, $(x - 2)$, or $(x + 2)$ are exact factors of $x^3 - 2x^2 - x + 2$.

69. $(x - 1)$, $(x + 1)$, $(x - 2)$, $(x + 2)$, $(x - 3)$, or $(x + 3)$ are exact factors of $x^4 + x^3 - 7x^2 + 6$.

In each of the following, find the quotient and remainder. Check either by substitution or multiplication.

70. $(5y + 2y^2 - 1) \div (2y - 3)$ **71.** $(c + c^3 - 2 - 3c^2) \div (c - 1)$

72. $(a^3 - 3) \div (a + 1)$ **73.** $(1 - x + x^3) \div (x - 2)$

Perform the indicated operations and simplify the result:

74. $\dfrac{(x + h)^2 - x^2}{h}$ **75.** $\dfrac{(x + h)^2 - (x + h) - (x^2 - x)}{h}$

76. $\dfrac{(x + h)^2 + 3(x + h) + 6 - (x^2 + 3x + 6)}{h}$

77. $\dfrac{2(x + h)^2 - 7(x + h) - 4 - (2x^2 - 7x - 4)}{h}$

CHAPTER 4

Simple Equations

A topic of considerable importance in algebra is one concerned with the equality of algebraic expressions. It is by means of equalities that generalizations can be made and problems can be solved. Our present interest is to distinguish between two important types of equalities and to introduce procedures which will, by a series of logical steps, direct us to particular solutions if they exist.

Types of Equalities

An equality is simply a statement that two expressions represent the same quantity. Thus $2x + 1 = x$ is an equality. So also is $2(x + 1) = 2x + 2$. The two expressions which are connected by the equal sign are called *members* of the equality—the *left member* is the entire expression to the left of the equal sign and the *right member* is the entire expression to the right of the equal sign. In the equality $2x + 1 = x$, the expression $2x + 1$ is the left member, and the monomial x is the right member.

Equalities are divided into two sets—*identical equations* or *identities*, which are true for all admissible values of the literal quantities, and *conditional equations* or *equations*, which do not hold true for all admissible values of the literal quantities. By *admissible values*, we mean values for which the expressions have meaning. For example, since division by zero is not possible, the expression $1/x$ does not have meaning for $x = 0$. Similarly, in the expression $3/(2x - 4)$, the value $x = 2$ is not an admissible value since this value will make the denominator zero.

65

The equal sign is used as a symbol of equality for both identities and equations. However, if it is desirable to emphasize that an equality is an identity, the symbol $\equiv$ (read "is identically equal to" or "identically equals") is used in place of the equal sign.

Identities

The equality $2(x + 1) = 2x + 2$ is true for all values of x since it is a statement of the distributive law. Therefore it is an identity, and all values of x are admissible. If we set $x = 5$, the equality becomes $2(5 + 1) = 2 \cdot 5 + 2$, and so on for each value of x we select.

The equality $y^2 - 9/y + 3 = y - 3$ is also an identity since, by actual division, the quotient of $y^2 - 9$ and $y + 3$ is $y - 3$. Note that the left member has no meaning for $y = -3$ since this value would make the denominator zero. Therefore $y = -3$ is not an admissible value. All other values of y are admissible.

The simplest way to show that an equality is an identity is to actually perform the operations indicated in one or both members until the two members are identical in form. As an illustration, consider the equality $3(2x - 4) = 2(3x + 5) - 22$. If we perform the indicated operations in the left member, it reduces to $6x - 12$. If we perform the indicated operations in the right member, it also reduces to $6x - 12$. Since the two expressions are identical in form, the equality is an identity.

Equations

The equality $2x + 1 = x$ does not hold true for all admissible values of x. For example, the equality is not true for $x = 3$. Note, however, that this is an admissible value of x since the expressions $2x + 1$ and x, which represent the members of the equality, have meaning for $x = 3$. Since this equality is false for at least one admissible value of x, it is classified as a conditional equation or simply an equation.

If an equation involves only one variable, a value of the variable for which the equation is true is called a *root* of the equation or a *solution* of the equation. To *solve* such an equation means to find all its roots or prove that it has none.

The solution of an equation requires a specification of the set of numbers under consideration. For example, if x is an element of the set of natural numbers, then the equation

$$2x + 1 = x$$

has no solution, since there is no natural number for which the equality is true. On the other hand, if x is an element of the set of integers, a solution of the equation $2x + 1 = x$ is the value $x = -1$.

Sometimes no specification of the number set to be used is given. If the equation arises from practical applications, the nature of the admissible values for the given problem will dictate the set of numbers which can reasonably be intended. Thus if x represents the length of a rectangle, the set of admissible values is restricted to positive numbers.

If an equation involves more than one variable, such as $2x + y = 13$, it can be satisfied by an unlimited number of combinations of specific values. Each such combination of values is called a solution of the equation. For example, the combination $x = 5$ and $y = 3$ is a solution since $2 \cdot 5 + 3 = 13$. So also are the combinations $x = 1$ and $y = 11$, $x = 10$ and $y = -7$, $x = -\frac{1}{2}$ and $y = 14$, and so on. It is important to recognize that although this equality is true for an unlimited number of values, it is not an identity since it is not true for *all* admissible values of x and y. For example, $x = 4$ and $y = 8$ are admissible values since the expression $2x + y$ has meaning for these values, but the equality is false for this combination of numbers. There are, of course, many other exceptions to the equality, but one exception is sufficient to identify it as an equation rather than an identity.

Exercise 13

EQUALITIES

Verify the following identities for $x = 4$, $x = -4$, $x = \frac{1}{2}$.

1. $5(2 + x) = 10 + 5x$ **2.** $3x = x + x + x$

3. $-4(x - 1) = -4x + 4$ **4.** $x(x + 7) = x^2 + 7x$

5. $(x - 1)(x - 2) = x^2 - 3x + 2$

6. $(2x + 3)(x + 4) = 2x^2 + 11x + 12$

7. $(x + 6)^2 = x^2 + 12x + 36$ **8.** $\dfrac{6x^2 + 13x - 5}{2x + 5} = 3x - 1$

List the nonadmissible values (if any) of each of the following identities:

9. $\dfrac{2x^2 + x}{x} = 2x + 1$ **10.** $\dfrac{x^4 - 1}{x^2 + 1} = x^2 - 1$

11. $3(x^2 - x) = 3x^2 - 3x$

12. $\dfrac{y^3 - 3y^2 + 2y}{y} = y^2 - 3y + 2$

13. $\dfrac{a^2 - 5a + 6}{a - 2} = a - 3$

14. $\dfrac{x^2 - y^2}{x - y} = x + y$

Show that the following equalities are identities by performing the indicated operations in each member:

15. $7(x^2 - 2x + 3) = 7x^2 - 14x + 21$

16. $4(x - 2) = x(4 - x) + x^2 - 8$

17. $3(x^2 - x + 5) = 3x^2 - 3(x - 5)$

18. $8 + 6(y - 3) = 2(3y - 5)$

19. $(a - 2)(a + 5) = a^2 + 3a - 10$

20. $(2c - 3)(3c - 2) = (3 - 2c)(2 - 3c)$

21. $(4v - 1)^2 = (1 - 4v)^2$

22. $(2x - 1)(4x - 3) = 3 + 2(4x^2 - 5x)$

23. $\dfrac{3x^2 + 5x - 2}{3x - 1} = x + 2$

24. $\dfrac{6x^2 + 13xy + 6y^2}{2x + 3y} = 3x + 2y$

Determine which of the following equalities are identities and which are equations.

25. $4x + 5 = 5x + 4$

26. $4x + 5 = 5 + 4x$

27. $3(x - 2) = 3(2 - x)$

28. $x(5 - 2x) = -x(2x - 5)$

29. $\dfrac{x^2 + 7x}{x} = x + 7$

30. $\dfrac{x^2 + 7x}{x} = x^2 + 7$

31. $(x + 3)(2x - 5) = 2x^2 - 15$

32. $(3x + 2)(3x - 2) = 9x^2 - 4$

33. $\dfrac{x^2 + 15x + 56}{x + 7} = x + 8$

34. $\dfrac{x^2 + 5x + 6}{x - 2} = x - 3$

35. $(12x + 8) \div 2 = 12x + (8 \div 2)$

36. $x = 9$

37. $x + y = 9$

38. $9(x + y) = 9x + 9y$

39. $(2x + y)^2 = 4x^2 + 4xy + y^2$

40. $3x - 2y = 2x - 3y$

41. $x + 2y = z$

42. $(x + 2)^2 = x^2 + 4$

The root of the equation $3x - 2 = 10$ is $x = 4$. Write the new equation, and show that the root does not change if each member of the given equation is:

43. Multiplied by 3.

44. Divided by 3.

45. Increased by 2.

46. Decreased by 2.

Determine, by a process of systematic trial, three solutions for each of the following equations:

47. $x + y = 10$ **48.** $x - y = 6$
49. $x + 2y = 15$ **50.** $2x - y = 4$
51. $3x + y = 9$ **52.** $3x - y = 9$
53. $2x + 3y = 24$ **54.** $4x - 3y = 12$

Which of the following equations are satisfied by the values $x = 4$, $y = -3$?

55. $3x + 2y = 6$ **56.** $3x + 2y = 18$
57. $2x - y = 5$ **58.** $5x - y = 23$
59. $x - 2y = 16$ **60.** $x = y + 7$
61. $x + 2y - 2 = 0$ **62.** $y = 2x - 11$

63. Is the formula $C = 2\pi r$ an identity or an equation? Explain your answer.

14   Solution of Simple Equations

Postulates of Equality

The solution of an equation is determined by performing a series of simplifying operations on its members which will change the form of the equation without changing the value of its roots. To understand the logical basis for performing such operations, it is necessary to formulate the properties which are assigned to all equalities. These properties are stated in the following set of postulates.

1. *The Reflexive Property.* $a = a$.
This property, which states that any number is equal to itself, may seem trivial, but it assures us that a given symbol retains its value throughout a discussion.

2. *The Symmetric Property. If $a = b$, then $b = a$.*
This property states that the two members of an equality may be interchanged. For example, if $5 = x + 3$, then $x + 3 = 5$. The use of this property sometimes simplifies the solution of an equation.

3. *The Transitive Property. If $a = b$, and $b = c$, then $a = c$.*
This property states that "Quantities equal to the same quantity are equal to each other." For example, if $x = y$ and $y = 4$, then $x = 4$.

4. *The Additive Property.* *If $a = b$, then $a + c = b + c$.*

This property states that "If equals are added to equals, the results are equal." For example, if $x - 5 = 7$, we may add 5 to both members of this equation, so that $x - 5 + 5 = 7 + 5$, or $x = 12$. We observe that the additive property implies that we may also subtract the same number from both members of an equation, since c may be a negative number.

5. *The Multiplicative Property.* *If $a = b$, then $ac = bc$.*

This property states that "If equals are multiplied by equals, the results are equal." For example, if $\frac{1}{2}x = 5$, then $\frac{1}{2}x(2) = 5(2)$, or $x = 10$. We observe that the multiplicative property implies that we may also divide both members of an equality by the same number, since c may be a fraction. Thus if $3x = 15$, we may divide both members of this equation by 3, so that $x = 5$.

Solution of a Simple Equation

Both the equalities

$$3(x - 5) - 3 = 2x - (7x + 2)$$
$$\text{and } x = 2$$

are equations. Each also has a single root with a value of 2. The central problem in the solution of simple equations is to transform an equation like the first, whose root is not apparent, to an equation like the second, whose root is obvious. In making the transformation, each step is justified by a postulate of equality or an algebraic operation based on previous postulates and definitions. A series of steps which will transform the first equation listed above into the second is illustrated in the following example.

EXAMPLE 1.   Solve the equation $3(x - 5) - 3 = 2x - (7x + 2)$.

$3(x - 5) - 3 = 2x - (7x + 2)$	Given equation
$3x - 15 - 3 = 2x - 7x - 2$	Apply distributive law to remove parentheses
$3x - 18 = -5x - 2$	Combine like terms
$8x - 18 = -2$	Add $5x$ to each member
$8x = 16$	Add 18 to each member
$x = 2$	Divide each member by 8

Check.   If we substitute $x = 2$ in the given equation, we have

$$3(2 - 5) - 3 = 2(2) - [7(2) + 2]$$
$$3(-3) - 3 = 4 - (14 + 2)$$
$$-9 - 3 = 4 - 16$$
$$-12 = -12$$

Since the given equation is satisfied for $x = 2$, we may be certain that $x = 2$ is the solution of the equation.

The check is considered an integral part of the solution: the value $x = 2$ represents a *necessary* value if the equality is to be true; the check verifies that this value is *sufficient* to make the equality true.

With practice, the steps which will effect a solution of an equation of this type will be direct and systematic.   In general, the procedure is as follows:

1. Remove parentheses by performing the indicated operations.
2. Combine similar terms occurring in the same member.
3. Add the negative of each variable term appearing in the right member to both members.   This will eliminate the variables from the right member and confine them to the left member.
4. Add the negative of each constant appearing in the left member to both members.   This will eliminate all such terms from the left member and confine them to the right member.   (Of course, it is possible to reverse the procedure so that all variables are confined to the right member and all constants to the left member, if this is more convenient).
5. Divide each member by the coefficient of the unknown.
6. Check the resulting solution.

The equations considered at present are such that each can be reduced to the form $ax + b = 0$ where x is the variable and a and b are constants. Such equations are called *first degree* or *linear equations* in one variable. They represent a large and important set of equations which are applicable to a wide variety of problem situations.

Exercise 14

Solution of Simple Equations

Each of the following statements is true because of one of the postulates of equality.   Identify the property of equality which justifies the statement.

1. If $-4 = x$, then $x = -4$.   2. If $6x = 18$, then $x = 3$.
3. If $y + 3 = 10$, then $y = 7$.   4. If $4x = y$ and $y = 8$, then $4x = 8$.
5. $2x = 2x$.   6. If $\frac{1}{2}a = 5$, then $a = 10$.

7. If $\frac{v}{2} = 7$, then $v = 14$.   8. If $8 = x - 6$, then $x - 6 = 8$.

9. If $x - 6 = 8$, then $x = 14$.   10. $5x - 4 = 5x - 4$.

11. If the reflexive, symmetric, and transitive properties of equalities were applied to the relationship between perpendicular lines, the resulting statements would be:

(*a*) *a* is perpendicular to *a*.   (reflexive property)

(*b*) If *a* is perpendicular to *b*, then *b* is perpendicular to *a*.   (symmetric property)

(*c*) If *a* is perpendicular to *b* and *b* is perpendicular to *c*, then *a* is perpendicular to *c*.   (transitive property)

Which of these three statements hold true for perpendicular lines in a plane?

Which of the following relations, as applied to the given elements, are reflexive, symmetric, or transitive?

12. "is parallel to," as applied to straight lines.

13. "is complementary to," as applied to angles.

14. "is in the same mathematics class as," as applied to students.

15. "is an exact factor of," as applied to algebraic expressions.

Simplify each of the following expressions:

16. $y - 3 + 3$ **17.** $x + 8 - 8$ **18.** $3v \div 3$

19. $(\frac{1}{2}b) \cdot 2$ **20.** $\frac{c}{4} \cdot 4$ **21.** $(-\frac{1}{5}x) \cdot (-5)$

In each of the following, determine an operation with a specific number which will reduce the given expression to x.

22. $x + 7$ **23.** $x - 2$ **24.** $3x$
25. $5 + x$ **26.** $-x$ **27.** $-6x$
28. $\frac{1}{4}x$ **29.** $\frac{2}{3}x$ **30.** $1\frac{1}{5}x$
31. $-\frac{3}{4}x$ **32.** $0.4x$ **33.** $-2.9x$

34. $x - 1$ **35.** $\frac{x}{8}$ **36.** $-\frac{2x}{5}$

Solve each of the following equations. Indicate accurately and completely the operation which is performed on each member of the equation. Check that the value obtained is an actual root.

37. $x + 4 = 12$

38. $x - 3 = 6$

39. $b + 5 = -6$

40. $z - 30 = -14$

41. $c + 27 = -10$

42. $3 + p = 5$

43. $3m = 45$

44. $7u = 42$

45. $2n = -18$

46. $-5v = 30$

47. $-8q = -12$

48. $\frac{1}{2}r = 28$

49. $\frac{1}{4}w = -5$

50. $-\frac{1}{3}u = 18$

51. $-\dfrac{x}{5} = -11$

52. $\frac{2}{3}s = 26$

53. $\frac{3}{5}t = -9$

54. $r - 3.4 = 1.7$

55. $0.1h = 10$

56. $4k = -13$

57. $\frac{1}{3}t = -8$

58. $\dfrac{c}{2} = -1$

59. $-y = 7$

60. $-x = -\frac{1}{2}$

Solve and check each of the following equations:

61. $3x + 4 = 13$

62. $5y + 7 = 22$

63. $2a - 3 = 9$

64. $4x - 8 = 28$

65. $6z - 14 = 16$

66. $3u - 10 = 14$

67. $3y + 5 = y + 17$

68. $4t + 3 = t + 18$

69. $8 = 11 + 3x$

70. $2 - x = 5x$

71. $4n = 6 + n$

72. $25 = 31 - 3v$

73. $14 - u = 4 - 3u$

74. $10r - 7 = 5r + 3$

75. $11 - x = 7x - 5$

76. $2b + 4 = -b - 7$

77. $y = 2y - 1$

78. $13c - 7 = 8 - 3c$

79. $h - 5 = 4h - 1$

80. $12 - 5n = 6n + 1$

81. $x - 3 + 2x = 7 - 2x$

82. $4(v - 2) = 6v$

83. $8z = 3(4 - 2z)$

84. $14t - 5 = 6(2t + 1) - 3$

85. $6(s - 4) + 3 = 5s$

86. $7 - 2(3y - 5) = 8y + 17$

87. $3 - w(4 + w) = 5 - w^2$

88. $3x^2 + 4x - 2 = 17 - 3x(5 - x)$

89. $4 - 2(t + 6) = 3t + 3(2 - t)$

90. $4(t - 5) + 8 = 3(2t - 4)$

91. $7(3x - 1) = 7(1 - 3x)$

92. $15 - (2 + 3v) = 9v + 1$

93. $4(y - 2) - 3(y + 2) = 12$

94. $3(2 - 3n) + 5(2n - 3) = 6 - n$

95. $(a + 7)(a - 7) = a(2 + a) - 1$

96. $(3w - 1)(2w - 3) = 6w^2 - (4 - 3w)$

97. $5 + (c - 3)(c + 2) = c^2 - 7c + 6$

98. $3x - (5 + x)(5 - 2x) = 2x(1 + x)$

99. $10 - 2(v - 6)^2 = v^2 - 3v(v - 6)$

100. $(2y + 3)^2 - 5(y + 2) = (4y - 2)(y + 1)$

In each of the following formulas, substitute the given values and solve the resulting equation for the required literal quantity:

101. $E = IR$.   (a) Find R if $E = 110$ volts, $I = 5.5$ amperes.

(b) Find I if $E = 114$ volts, $R = 38$ ohms.

(c) Find E if $I = 4$ amperes, $R = 0.5$ ohm.

102. $p = 2(l + w)$.   (a) Find w if $p = 48$ feet, $l = 5$ feet.

(b) Find p if $l = 3\frac{1}{2}$ inches, $w = 2\frac{1}{3}$ inches.

103. $s = \frac{1}{2}at^2$.   Find a if $s = 360$ feet, $t = 6$ seconds.

104. $A = p + prt$.   (a) Find r if $p = \$600$, $t = 2$ years, $A = \$648$.

(b) Find p if $A = \$515$, $r = 0.06$, $t = \frac{1}{2}$ year.

105. $d = rt$.   (a) Find t if $d = 180$ miles, $r = 40$ miles per hour.

(b) Find r if $d = 44.8$ centimeters, $t = 0.04$ second.

106. $C = 2\pi r$.   Find r if $C = 121$ inches.   Use $\pi = \frac{22}{7}$.

107. $A = \frac{1}{2}h(b + b')$.   (a) Find h if $A = 68$ square inches, $b = 10$ inches, $b' = 7$ inches.

(b) Find b' if $A = 5.6$ square feet, $h = 0.8$ foot, $b = 9$ feet.

108. $P = 4s$.   Find s if $P = 228$ yards.

109. $V = \frac{1}{3}Bh$.   (a) Find B if $V = 76$ cubic feet, $h = 19$ feet.

(b) Find h if $V = 7\frac{1}{2}$ cubic inches, $B = 3\frac{1}{2}$ square inches.

15   Expressing Problems in Symbolic Form

One of the critical steps in solving a verbal problem is translating the relations of the problem into algebraic symbols.   This requires first an ability to express unknown quantities in terms of literal numbers, and second, an ability to form equations which involve the literal quantities. Because of their role in applied problems, the literal numbers in such equations are frequently referred to as *unknowns*.   Our first concern is to form algebraic expressions which will embody the properties assigned to an unknown in a problem situation.

Forming Algebraic Expressions

In forming an algebraic expression representing an unknown quantity, it may often be helpful to invent a parallel situation involving specific numbers.   For example, the expression which represents the amount by which n exceeds 5 can be obtained by considering the expression which represents the amount by which 12 exceeds 5. Since 12 exceeds 5 by $12 - 5$ (or 7), n exceeds 5 by $n - 5$.

EXAMPLE 1. What is the next even integer after $2n$? The next even integer after 8 is 10, which is $8 + 2$. The next even integer after 26 is $26 + 2$ or 28. Thus, the next even integer after $2n$ is $2n + 2$.

EXAMPLE 2. How many cents are there in D dimes and N nickels? In 3 dimes, there are $3 \cdot 10$ or 30 cents; in 7 nickels, there are $7 \cdot 5$ or 35 cents. Similarly, in D dimes, there are $D \cdot 10$ or $10D$ cents; in N nickels, there are $N \cdot 5$ or $5N$ cents. Therefore in D dimes and N nickels there are $10D + 5N$ cents.

EXAMPLE 3. If Bob is y years of age, how old was he n years ago? Three years ago, he was $y - 3$ years of age; 5 years ago, he was $y - 5$ years old. Therefore n years ago, Bob was $y - n$ years of age.

EXAMPLE 4. If a wheel makes R revolutions in t minutes, how many revolutions will it make in 1 hour? If it were to make 100 revolutions in 5 minutes, it would make 100/5 or 20 revolutions per minute. Therefore, if it makes R revolutions in t minutes, it makes R/t revolutions per minute. Since there are 60 minutes in one hour, it would make $60R/t$ revolutions in one hour.

Forming Algebraic Equalities

In translating a verbal statement into the form of an algebraic equality, it is necessary to find some quantity in the statement which can be expressed in two ways. In the most elementary situations, the equality is expressed directly and needs only to be written in terms of algebraic symbols.

EXAMPLE 5. Write the equality which represents the statement: "When a certain number is increased by 15, its value is doubled." If we let n represent the number, the equality becomes

$$n + 15 = 2n$$

where the left member indicates the number is increased by 15, the right member indicates the number is doubled, and the equal sign shows that these expressions represent the same quantity.

When the equality is not stated directly, the relations are more subtle. No general procedure can be given, but the following examples and those in the next section illustrate the method in some common situations.

EXAMPLE 6. What equality is equivalent to the statement: "The larger of two numbers, whose sum is 18, exceeds the smaller by 4"? Since the sum of the two numbers is 18, if we let

$$x = \text{the smaller number}$$
$$\text{then } (18 - x) = \text{the larger number}$$

The phrase "the larger exceeds the smaller by 4" can be interpreted in three ways, all of which are equivalent.

1. If the smaller is subtracted from the larger, the difference is 4. This leads to the equation

$$(18 - x) - x = 4$$

2. If the smaller is increased by 4, it will equal the larger. This leads to the equation

$$x + 4 = 18 - x$$

3. If the larger is decreased by 4, it will equal the smaller. This leads to the equation

$$(18 - x) - 4 = x$$

These three equations are equivalent, that is, they have the same root. Any one equation is sufficient to represent the verbal statement.

EXAMPLE 7.   Write the equation for the following verbal statement: "The total value of 300 tickets, some at $1.80 each and the rest at $2.40 each, is $660." Let x = number of tickets at $1.80 each. Then, $300 - x$ = number of tickets at $2.40 each. The value, in dollars, of x tickets at $1.80 each is $1.8x$ and of $(300 - x)$ tickets at $2.40 each is $2.4(300 - x)$. The required equation is

$$1.8x + 2.4(300 - x) = 660$$

Exercise 15

EXPRESSING PROBLEMS IN SYMBOLIC FORM

Express each of the following in algebraic symbols:

1. Six more than x.
2. y decreased by 7.

3. The quantity which is n greater than m.

4. The difference between b and 8, in the order stated. *b - 8*

5. c less than u.

6. The sum of 4 and twice x.

7. The number which is 10% greater than y. *1.1y*

8. What number exceeds n by 9? *9+n*

9. What number exceeds n by a? *n+a*

10. What is the next consecutive odd integer after $2n + 1$?

11. If the sum of two numbers is 17 and one of the numbers is x, what is the other number? *17 - x*

12. The smaller of two numbers, whose difference is 9, is r.   What is the other number?

13. What is the cost of 14 books at x dollars each?

14. Two numbers total N.   If the smaller is b, what is the larger? *N - b*

15. How many minutes are there in h hours? *60h*

16. What is the total value, in cents, of q quarters and d dimes? *25q + 10d*

17. Find the combined value of m 4-cent stamps and n 5-cent stamps. *4m + 5n*

18. John is n years old.   How old will he be in 6 years?

19. John is n years old, and his father is three times as old as John was 3 years ago.   How old is the father now? *3(n-3)*

20. John is n years old.   In 6 years, his father will be twice as old as John will be.   How old is his father now? *2(n+6)-6*

21. How far will a car travel in 3 hours at r miles per hour?

22. Two motorists travel in opposite directions from the same point. If one travels at r_1 miles per hour and the other at r_2 miles per hour, how far apart will they be after t hours? *$r_1 t + r_2 t$*

23. A carpenter can nail n lineal feet of flooring in h hours.   How many feet of flooring can he nail in 1 minute? *$\frac{n}{60h}$*

24. What is the simple interest on $1200 at r per cent for n years?

25. The width of a rectangle is w.   If the length is 4 more than twice the width, what is the perimeter?

Write an equation for each of the following statements:

26. The smaller of two numbers is n, the larger is $3n$, and
    (a) the sum of the two numbers is 80. *n + 3n = 80*
    (b) the larger number is 9 greater than the smaller. *3n + 9 - n*
    (c) the larger number exceeds twice the smaller by 26. *3n - 26 - n*
    (d) the sum of the two numbers is 19 less than twice their difference.

$n + 3n$
$2(3n - n) - 19$

27. The sum of two numbers is 24, the smaller is x, and

    (a) the larger exceeds the smaller by 18.

    (b) the sum of the larger and four times the smaller is 45.

    (c) their difference is 9 less than the smaller number.

28. The smallest of three consecutive integers is y, and

    (a) the sum of the three consecutive integers is 96.

    (b) twice the smallest is 3 less than the sum of the remaining two integers.

    (c) three times the largest integer exceeds the smallest by 22.

29. In a collection of coins, there are d dimes and $(22 - d)$ quarters, and

    (a) the total value of the collection is \$3.85.

    (b) the number of dimes is 8 more than the number of quarters.

    (c) the value of the dimes is \$1.70 more than the value of the quarters.

30. An eastbound train leaves town A at an average speed of r miles per hour at the same time a westbound train leaves the same town traveling at an average speed of $(r + 10)$ miles per hour, and

    (a) the two trains are 300 miles apart at the end of 2 hours.

    (b) the distance between them is increasing at an average rate of 140 miles per hour.

    (c) the speed of the westbound train is 45 miles per hour less than twice the speed of the eastbound train.

16   Solution of Problems

Applied problems which can be solved by the use of algebra are so varied that no single procedure can cover all variations.   There are, however, certain steps that will give direction to the analysis of a particular problem.   They are listed as they apply to several illustrative examples.

EXAMPLE 1.   A man is 26 years older than his son.   Six years ago, he was three times as old.   How old is the son?

Step 1.   Read the problem for understanding.   This may mean reading it several times, visualizing it, experimenting with specific numbers, drawing a diagram, or estimating an answer.   In particular, it is necessary to identify the facts given in the problem and to determine what is required.   The facts given in this problem are: (a) at present, the father's age is 26 years greater than the son's age, and (b) when the age of each was 6 years less, the father was 3 times as old as the son.   It is required to determine the son's present age.

Step 2.   Represent each unknown quantity by a literal number or an expression containing the literal number.   The literal expression must represent some measurable quantity.   Describe it accurately.   In the stated problem, there are two unknowns, the father's age and the son's age.   We can choose to let

$$x = \text{son's age, in years, at the present time}$$
$$\text{then } x + 26 = \text{father's age, in years, at present time}$$

Step 3.   Write an equality based on the facts of the problem.   If the equality is an identity, discard it and form another which is an equation and involves the expressions representing unknown quantities.   For the problem under consideration, the fact that 6 years ago the father's age was 3 times the son's age leads to an equality.   Six years ago, the father's age was $(x + 26) - 6$ or $(x + 20)$.   Therefore

$$x + 20 = 3(x - 6)$$

Since this is an equation and not an identity, it can be used to solve the problem.

Step 4.   Solve the equation and use the solution to evaluate each unknown required by the problem.   The solution of the equation developed in Step 3 is:

$$x + 20 = 3(x - 6)$$
$$x + 20 = 3x - 18$$
$$-2x = -38$$
$$x = 19$$

Therefore the son's present age is 19 years.

Step 5.   Check the results by substituting in the original problem. From the results obtained in Step 5,

$$x = \text{son's age now} = 19 \text{ years}$$
$$\text{then } x + 26 = \text{father's age now} = 45 \text{ years}$$

Therefore the father is 26 years older than the son.   Six years ago, the son was 13 and the father was 39.   Therefore the father was three times as old as the son.

EXAMPLE 2.   What is the distance between two towns if a train which requires 5 hours to travel between them could reduce its running time 1 hour by increasing its speed 15 miles per hour?

Step 1. Understanding this problem involves ability to visualize two trips of the train, one at its usual speed and the other at its increased speed. This is what is known:

Trip 1: Town *A*   unknown distance   Town *B*

unknown speed
time = 5 hours

Trip 2: Town *A*   unknown distance   Town *B*

speed = initial speed + 15
time = 4 hours

Step 2. We have a choice of selecting x to represent the unknown distance or the unknown initial speed. Either is suitable although it is simpler to make the following representation since it will avoid fractions.

Let x = initial speed of the train in miles per hour
then, since $d = rt$ (distance equals rate times time)
$5x$ = distance between the towns, in miles

It is now possible to list the information in the form of a table as follows:

Trip	Rate	Time	Distance
1	x	5	$5x$
2	$x + 15$	4	$5x$

Step 3. The distance between the two towns can be expressed in several ways. It is represented by the expression $5x$. It can also be expressed as the product of rate and time since

$$d = rt$$

For trip 1, this leads to the identity

$$5 \cdot x = 5x$$

For trip 2, it leads to the equation

$$4(x + 15) = 5x$$

Step 4. Solving the equation, we have

$$4(x + 15) = 5x$$
$$4x + 60 = 5x$$
$$x = 60$$

Although $x = 60$ is the solution of the equation, it is not the solution of the problem since x represents the initial speed of the train, and we are required to find the distance between the two towns. However, we know

$$x = \text{initial speed of the train} = 60 \text{ miles per hour}$$
$$\text{and } 5x = \text{distance between towns} = 300 \text{ miles}$$

Therefore the required distance is 300 miles.

Step 5.   Check: If the distance between towns is 300 miles, a train running at 60 miles per hour would require 5 hours, as stated in the problem. If the speed is increased to 75 miles per hour, the train would then travel the 300 miles in 4 hours.

Exercise 16

SOLUTION OF PROBLEMS

1. A length of steel pipe is 15 feet long. It is to be cut into two pieces. If one piece is to be 4 feet longer than the other, how long should each piece be?

2. The top of a table is to be a rectangle such that the length is 2 feet more than the width. A chrome molding is to be nailed around the outside of the table top. Find the dimensions of the table top which will require 18 feet of molding.

3. Find two consecutive odd numbers whose sum is 56.

4. The sum of two numbers is 69.   The larger exceeds twice the smaller by 6.   Find the numbers.

5. In a triangle, the largest angle is three times the smallest and 9° greater than the third angle.   Find the number of degrees in each angle (the sum of the 3 angles of a triangle equals 180°).

6. Six men intend to purchase a factory, sharing the cost equally.   By including one additional man in the venture, the cost to each will be decreased $740.   What is the cost of the factory?

7. If a number is increased by 3, its square will be increased by 111. Find the number.

8. A father is 26 years older than his son.   In 9 years, he will be twice as old as his son.   Find their ages now.

9. The smaller of two numbers is one-third the larger, and their sum is 56.   Find the numbers.

10. If the opposite side of a square are each increased by 3 inches, and the remaining pair of sides are each decreased by 2 inches, the area of the figure does not change.   What is the length of each side of the square?

11. Tom has $5.20 in quarters and dimes.   He has 3 more dimes than quarters.   How many of each does he have?

12. One number is twice a second number.   Eight more than the first number is the same as the second number less 9.   Find the numbers.

13. Two employees have an aggregate of 51 years experience with their firm.   If one of them has been employed 5 years longer than the other, how many years has each been with the firm?

14. How fast does a car travel if in 2 hours it overtakes a bus which averages 36 miles per hour and leaves 1 hour before the car?

15. Harry is twice as old as his brother.   Four years ago, he was 4 times as old.   How old is each?

16. The sum of three consecutive even integers is 228.   Find the integers.

17. A rectangular garden is 15 feet longer than it is wide.   Find its dimensions if its perimeter is 118 feet.

18. The paid attendance at a football game was 5130.   Some paid $2 for their tickets while the rest paid $1.   If the total receipts were $8170, how many tickets of each kind were sold?

19. Two planes leave an airport at the same time but travel in opposite directions.   If the first plane travels at a speed of 280 miles per hour and the second travels 15 miles per hour faster, in how many hours will they be 690 miles apart?

20. A large office employs 200 workers, with a daily payroll of $3520. If those without secretarial skills earn $14 per day, and those with secretarial skills earn $20 per day, find the number employed in each classification.

21. The base of a triangle is 5 inches longer than one side and twice as long as the remaining side.   The perimeter of the triangle is 50 inches. Find the length of each side.

22. Twenty-seven coins consist of quarters, dimes, and nickels.   If there is one more nickel than quarters, and four more dimes than nickels, find the number of each kind of coin.

23. In 20 years, Ann will be three times as old as she was 6 years ago. What is her present age?

24. In financing a road costing $1,960,000, the county contributed 30% more than the city, and the state contributed twice as much as the county.   Find the amount contributed by each.

25. Bill has an equal number of $1, $5, and $10 bills.   He has $112 in all.   How many bills of each denomination does he have?

26. The sum of four angles about a point is 360°.   The second is three times the first, the third is twice the second, and the fourth is 40° greater than the third.   What is the size of each angle in degrees?

27. A man has $6000 invested, part at 5% and the balance at 4%.   If his total income from the two investments is $275 per year, how much does he have invested at each rate?

28. An office requisition calls for 300 stamps, some 4-cent and some 5-cent, for $13.   How many of each kind were requested?

29. The sound made by a workman spiking down a rail in a railroad track comes to an observer 1.2 seconds sooner through the rails than through the air.   Assuming that sound travels through the air at 1100 feet per second, and through the rails at 1700 feet per second, how far from the observer is the workman?

30. A triangular lot has a perimeter of 142 feet.   Is it possible that the lengths of the 3 sides are consecutive integers?   Explain your answer.

17   Inequalities

In Section 16 we solved problems which could be expressed by simple equations; the solution of the equation was then the solution of the problem.   There are numerous problems, however, which require the solution of an *inequality*.   Let us consider one such problem.

> A salesman has a choice of two jobs.   In the first job his salary will be $100 per week plus a bonus of 20% of his sales during the week.   In the second job his salary will be $150 per week plus a bonus of 10% of his sales during the week.   He realizes that when his sales are low the second job will pay him more than the first job, and that when his sales are high the first job will pay him more than the second.   He would like to know what his weekly sales must be in order to earn more from the first job than from the second.

If we let s equal the sales for one week, and i the income for that week, then we shall have for the first job

$$i = 100 + 0.20s$$

For the second job we have

$$i = 150 + 0.10s$$

The condition that the first job pay more than the second is expressed by the *inequality*

$$100 + 0.20s > 150 + 0.10s$$

To answer the salesman's question, we must solve this inequality, that is, we must determine the values of s which will satisfy the inequality. Note that we have used the expression "values of s," indicating that inequalities, unlike simple equations, are satisfied by more than a single value of the variable. By methods to be discussed in a later part of this section we find that the solution of the inequality is $s > 500$. Therefore whenever the salesman's total sales for the week exceed \$500, the first job will pay him more than the second. That this solution is correct may be verified by comparing the total income from each job for sales of less than \$500 and sales of more than \$500. We may also note that when the sales total exactly \$500, the income from each job is the same.

The number scale, as in Figure 1, Section 4, may be used to indicate the range of values of the variable which satisfy an inequality. In the inequality discussed in the previous paragraph, the solution $x > 500$ is shown on the number scale of Figure 1. The small circle at $x = 500$

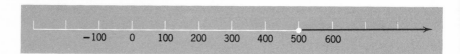

Figure 1

indicates that the value 500 is not included in the inequality $x > 500$. When the inequality is of the form $x \geq 8$, for example, we replace the circle with a dot, to indicate that $x = 8$ is included in the solution, as in Figure 2.

Figure 2

Types of Inequalities

Analogous to the two types of equalities (identities and equations of condition) are the two types of inequalities, absolute inequalities and conditional inequalities. Absolute inequalities are true for all admissible values of the variable, and conditional inequalities are not true for all values of the variable. The inequality

$$(x - 2)^2 > -1$$

is true for all admissible values of x, and hence it is an example of an absolute inequality. For if $x = 2$, $(x - 2)^2 = 0$, which is greater than -1, and for all other admissible* values of x, $(x - 2)^2$ is a positive number, and therefore greater than -1. The inequality

$$2x - 3 > 5$$

is a conditional inequality, since it is not true for all values of x. If $x = 2$, $2x - 3 = 1$, which is not greater than 5. But if x is any number greater than 4, $2x - 3$ is greater than 5, so that the inequality is satisfied for all values of x, and only values of x, which are greater than 4.

Postulates of Inequality

As in the case of equations, the solution of inequalities depends on certain postulates, some of which are quite similar to the postulates of equality. We state these postulates here.

1. *The Transitive Property*
   (a) If $x > y$, and $y > z$, then $x > z$.
   (b) If $x < y$, and $y < z$, then $x < z$.
Thus since $8 > 5$, and $5 > 3$, it follows from (a) that $8 > 3$. Also, since $4 < 6$, and $6 < 9$, it follows from (b) that $4 < 9$.

2. *The Additive Property*
   (a) If $x > y$, then $x + z > y + z$.
   (b) If $x < y$, then $x + z < y + z$.
Parts (a) and (b) of the additive property are frequently combined in the statement, "If equals are added to unequals, the results are unequal

---

* For certain values of x, called imaginary, or complex, numbers, $(x - 2)^2$ is not a positive number and must be excluded as solutions of the inequality. Such numbers are not included in this discussion.

in the same order." The expression "same order" means that both inequalities are "greater than," or that both inequalities are "less than."*
We may note that the additive property implies that we may also subtract the same number from both members of an inequality, since z may be a negative number.

Thus if we add 2 to both members of the inequality $6 > 4$, we have $6 + 2 > 4 + 2$, or $8 > 6$. If we subtract 2 from both members of the same inequality, we have $6 - 2 > 4 - 2$, or $4 > 2$.

3. *The Multiplicative Property*
  (a) If $x > y$, and z is a positive number, then $xz > yz$.
  (b) If $x > y$, and z is a negative number, then $xz < yz$.
  (c) If $x < y$, and z is a positive number, then $xz < yz$.
  (d) If $x < y$, and z is a negative number, then $xz > yz$.

The multiplicative property may be stated as "If unequals are multiplied by the same positive number, the results are unequals in the same order, and if unequals are multiplied by the same negative number, the results are unequals in the opposite order." Thus if both members of the inequality $6 > -3$ are multiplied by 2, the result is $12 > -6$, but if this inequality is multiplied by -2, the result is $-12 < 6$, thus reversing the sense of the inequality. We note that since multiplication by $1/n$ is equivalent to division by n, this property can be extended to the operation of division.

Solution of a Simple Inequality

We are now ready to apply the postulates of inequality to the solution of simple inequalities in much the same manner that we applied the postulates of equality to the solution of equations. We shall give several examples.

EXAMPLE 1.   Solve the inequality $x + 3 > -2$.

$$x + 3 > -2 \qquad \text{The given inequality}$$
$$x + 3 - 3 > -2 - 3 \qquad \text{Additive property}$$
$$x > -5 \qquad \text{Simplifying each member}$$

Check: For $x = -5$, $x + 3 = -2$. When $x > -5$, we increase the left member, while the right member remains -2. Therefore for all values of x greater than -5, $x + 3 > -2$.

---

* The expression "same sense" is frequently used instead of "same order."

In Figure 3, we indicate on the number scale the values of x for which this inequality is satisfied.

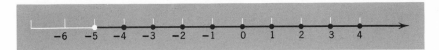

Figure 3

EXAMPLE 2.   Solve the inequality $3x + 5 \leq 2x + 11$.

$$3x + 5 \leq 2x + 11 \qquad \text{The given inequality}$$
$$3x - 2x + 5 - 5 \leq 2x - 2x + 11 - 5 \qquad \text{Additive property}$$
$$x \leq 6 \qquad \text{Simplifying each member}$$

Check: We may verify that the solution is correct by noting that when $x = 6$, both members of the inequality are equal; that when $x < 6$, the inequality is satisfied, that is, the left member is less than the right member.   When $x > 6$, however, the inequality is not satisfied, since the left member is then greater than the right member.

In Figure 4, we indicate on the number scale the values of x which satisfy the inequality.   Note the use of the dot, rather than the circle, to indicate that $x = 6$ is included in the values of x which satisfy the inequality.

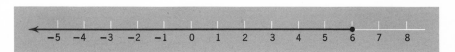

Figure 4

EXAMPLE 3.   Solve the inequality $6x + 7 > 8x - 3$.

$$6x + 7 > 8x - 3 \qquad \text{The given inequality}$$
$$-2x > -10 \qquad \text{Additive property}$$
$$x < 5 \qquad \text{Multiplicative property}$$

Note that the sense of the inequality has been changed, since we divided both members by -2.   The student may verify, as in the other examples, that the solution $x < 5$ is correct.   In Figure 5, we indicate on the number scale the values of x which satisfy the inequality.

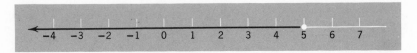

Figure 5

Solution of a Pair of Inequalities

Sometimes a variable must satisfy a pair of inequalities.   The following problem, for example, requires the solution of two inequalities.

The posted speed limits on a certain expressway are a maximum speed of 60 miles per hour and a minimum speed of 40 miles per hour.   A man driving home on this expressway telephones his wife from a telephone booth 240 miles from his home.   If the call is made at 2:00 P.M., between what hours can she expect him to arrive, assuming that he drives without stopping?

From the formula

$$\frac{\text{distance}}{\text{time}} = \text{rate}$$

we have the two inequalities

$$\frac{240}{t} \leq 60$$

$$\frac{240}{t} \geq 40$$

The first inequality is equivalent to

$$t \geq \frac{240}{60}, \text{ or } t \geq 4$$

The second inequality is equivalent to

$$t \leq \frac{240}{40}, \text{ or } t \leq 6$$

Therefore his travel time is between four and six hours, so he may be expected to arrive home between 6 P.M. and 8 P.M. that evening.   Figure 6 indicates the values of t which satisfy both inequalities.

Figure 6

The first and second inequalities may be combined in one expression as

$$4 \leq t \leq 6$$

which may be read, "t is greater than or equal to 4 and less than or equal to 6."

Exercise 17

Solution of Simple Inequalities

Given $a > b$, place either the symbol $>$ or $<$ between each of the following pairs of numbers:

1. $2a \quad 2b$ **2.** $a + 4 \quad b + 4$ **3.** $a - 2 \quad b - 2$

4. $-2a \quad -2b$ **5.** $-a \quad -b$ **6.** $\dfrac{a}{-2} \quad \dfrac{b}{-2}$

7. $2 - a \quad 2 - b$ **8.** $-2 - a \quad -2 - b$

Indicate the values of x on a number line for each of the following inequalities:

9. $x > 5$ **10.** $x < 5$ **11.** $x \geq -4$ **12.** $x \leq -1$
13. $2 < x < 8$ **14.** $2 \leq x \leq 8$ **15.** $-8 \leq x \leq -3$ **16.** $-2 \leq x \leq 5$

Indicate whether each of the following statements is true or false, giving a reason for your answer:

17. If $-x < 6$, then $x < -6$. **18.** If $-x < 0$, then x is negative.
19. If $-5x < -10$, then $x > 2$. **20.** If $-x < 5$, then $x < -5$.

State which of the following inequalities are absolute and which are conditional:

21. $x + 7 > 13$ **22.** $x + 2 > x + 1$ **23.** $2x - 4 > 8$

24. $(x - 3)^2 > 1$ **25.** $(x - 4)^2 > -1$ **26.** $(x - 2)^3 > -1$

27. $x^2 + 4 > 0$.

28. $2x > x$, where x is a positive integer.

29. $2x > x$, where x is a positive or negative integer.

30. $2x < x$, where x is a negative integer.

Solve the following inequalities, indicating the solution on a number line.

31. $2x - 3 < 7$ **32.** $5x - 4 < -12$

33. $2x + 5 > 4x - 9$ **34.** $3x - 5 \leq 2x + 12$

35. $4x - 3 \geq 21$ **36.** $x - 7 > 3x + 10$

37. $3x + 2 > x - 10$ **38.** $\dfrac{x}{3} + 5 \geq 7$

39. $\dfrac{x}{2} + 1 > x - 14$ **40.** $6x - 11 < 13$

41. $5 - 2x \geq 17 + 2x$ **42.** $\dfrac{2p}{3} - 1 > -3 + p$

Solve each of the following pairs of inequalities, indicating the solution on a number line:

43. $2x - 1 > 7$ **44.** $x - 1 \leq 5$ **45.** $3x - 1 \geq x + 7$
$3x - 1 < 17$ $x + 1 \geq 9$ $4x - 1 \leq 2x + 11$

46. $3x - 1 < 5x - 17$ **47.** $3x - 1 > 14$ **48.** $4 + x < -3$
$2x - 1 > x - 9$ $2x - 1 < -11$ $-2 + x > -9$

49. Solve the illustrative problem stated at the beginning of Section 17.

50. Mr. Smith's salary is $100.00 per week. He has a choice of two income protection plans which assure him a certain income when he is unable to work on account of illness. Plan A pays his full salary after four weeks of absence from work. Plan B pays 80% of his salary starting from the first day of absence from work. Both plans pay up to one year of illness. After how many weeks of absence will plan A pay him more total income than plan B?

18   Equations of the First Degree in Two Unknowns

Some of the verbal problems considered to this point involved more than one unknown quantity. In each case, however, we have been

able to express all unknowns in terms of one of them.   This may sometimes be difficult or inconvenient.   Therefore we now extend our consideration of equalities to the use of two first degree equations in two
unknowns.

Solution of a System of Equations

A pair of equations of the form

$$ax + by = c$$
$$dx + ey = f$$

where x and y represent variables and the remaining literal symbols
represent constants form a system of two linear or first degree equations
in two unknowns.   Each equation has an unlimited number of solutions,
that is, pairs of values of x and y which satisfy the equation.   We are
interested particularly in values of x and y which satisfy both equations.
Such a pair of values is called a *solution of the system of equations*.   A solution, if it exists, can be found by using the postulates of equality to
eliminate one of the unknowns.   Two methods of elimination will be
considered: elimination by addition and elimination by substitution.

Elimination by Addition

Consider the system

$$4x + 3y = 26$$
$$3x - y = 13$$

If we rewrite the first equation and multiply both members of the
second by 3, the coefficients of y will be numerically equal but opposite
in sign:

$$4x + 3y = 26$$
$$9x - 3y = 39$$

Now, recalling that if equals are added to equals, the sums are equal,
we add the left members and the right members.   The resulting equation
eliminates the term in y and becomes

$$13x = 65$$
$$\text{or } x = 5$$

This indicates that if there is a common solution, the necessary value of
x is 5.   To find the corresponding value of y, we can repeat the procedure

to eliminate x from the system, or we can substitute the value of x in either of the two given equations. Thus, substituting $x = 5$ in the first equation, we have

$$4x + 3y = 26$$
$$20 + 3y = 26$$
$$3y = 6$$
$$y = 2$$

Therefore the necessary values of x and y are $x = 5$, $y = 2$.

Check: For the first equation, when we substitute $x = 5$ and $y = 2$, we have

$$4(5) + 3(2) = 26$$
$$26 = 26$$

For the second equation, when we substitute $x = 5$ and $y = 2$, we have

$$3(5) - 2 = 13$$
$$13 = 13$$

Since both equations are satisfied by the values $x = 5$, and $y = 2$, $x = 2$ and $y = 5$ is the solution of the given system of equations.

The steps for elimination by addition are summarized as follows:

1. Multiply one or both equations by suitable constants so that the coefficients of one unknown are equal in absolute value but opposite in sign.

2. Add the two equations to eliminate the unknown.

3. Solve the resulting equation.

4. Find the value of the remaining unknown by repeating the procedure or by substituting in either equation.

5. Check by substituting in the original pair of equations.

Elimination by Substitution

Consider again the system

$$4x + 3y = 26$$
$$3x - y = 13$$

To solve this system by substitution, first solve either equation for one unknown in terms of the other. Since the coefficient of y in the second equation is -1, we can avoid fractions by solving for y in that equation.

Therefore

$$3x - y = 13$$
$$-y = 13 - 3x$$
$$y = 3x - 13$$

Substituting this value of y in the first equation eliminates the unknown y and enables us to solve for x:

$$4x + 3y = 26$$
$$4x + 3(3x - 13) = 26$$
$$4x + 9x - 39 = 26$$
$$13x = 65$$
$$x = 5$$

The value of y in terms of x has previously been found to be

$$y = 3x - 13$$

Substituting $x = 5$ in this equation, we have

$$y = 15 - 13$$
$$y = 2$$

Therefore the solution is $x = 5$, $y = 2$. The check is performed as before.

The steps for elimination by substitution are summarized as follows:

1. Solve either equation for one unknown in terms of the other (if possible, select an unknown whose coefficient is $+1$ or -1 to avoid fractions).
2. Substitute the expression obtained in the remaining equation.
3. Solve the resulting equation.
4. Find the value of the remaining unknown by substitution.
5. Check by substituting in the two original equations.

Consistent and Inconsistent Systems of Equations

The pair of equations

$$4x + 3y = 26$$
$$3x - y = 13$$

have one and only one solution, namely $x = 5$, $y = 2$ (this will be demonstrated in Section 38). Such a set of equations is called *consistent and independent.*

A system of equations may have no solution.   For example,

$$3x + 2y = 6$$
$$6x + 4y = 15$$

can have no solution, as is evident if we multiply each member of the first equation by 2 to obtain the equivalent system

$$6x + 4y = 12$$
$$6x + 4y = 15$$

Now, since each of the terms $6x$ and $4y$ represent a numerical quantity, it is impossible that their sum be both 12 and 15.   Therefore there is no set of values which can satisfy both equations.   Such a system of equations is said to be *inconsistent*.

It is also possible that a system of equations may have an unlimited number of solutions.   For example, the system

$$5x - 2y = 7$$
$$15x - 6y = 21$$

has an unlimited number of solutions.   This is evident if we multiply each member of the first equation by 3 to obtain the equivalent system

$$15x - 6y = 21$$
$$15x - 6y = 21$$

Since these two equations are identical, every solution of one equation is also a solution of the other.   Such a system of equations is said to be *consistent and dependent*.

Exercise 18

EQUATIONS OF FIRST DEGREE IN TWO UNKNOWNS

Solve each pair of equations by the method of elimination by addition:

1. $2x + y = 13$
   $2x - y = 11$

2. $5x + 4y = 7$
   $3x - 4y = -47$

3. $7x - 9y = 13$
   $x + y = -5$

4. $3x - 8y = -5$
   $x - 2y = -3$

5. $2x + 3y = -3$
   $3x - 2y = 54$

6. $5x - 4y = 2$
   $6x - 5y = 1$

7. $x + 3y - 8 = 0$
   $5x + 6 = y$

8. $3y - 2x = 7$
   $7 + 7x = 4y$

9. $y = 5x - 2$
   $x = 3y + 1$

Solve by the method of elimination by substitution:

10. $x + y = 1$
$\quad y = 2x - 11$

11. $3x + 2y = 5$
$\quad x = 5y - 4$

12. $5x + 6 = y$
$\quad y + 2 = x$

13. $6x - 5y = 10$
$\quad 2y - 3 = x$

14. $y = 2x + 23$
$\quad y = x + 13$

15. $x = 2y - 11$
$\quad 5x - 3y = 8$

16. $9x - 2y = 17$
$\quad x + 3y = -11$

17. $x + 3y = 15$
$\quad 3x - 7y = 3$

18. $3x - 5y = 13$
$\quad 7x - y = 25$

Solve each of the following.   Use either addition or substitution to eliminate one unknown.

19. $5a - 6d = -1$
$\quad 3a + 6d = 9$

20. $10m - 6n = 10$
$\quad 12m - 9n = 21$

21. $u = -3v$
$\quad 6u + 5v = 18$

22. $3r = 2s - 10$
$\quad s = 1 + r$

23. $6 = 3b - 7c$
$\quad 2 = 6c - 4b$

24. $x = 6 - y$
$\quad 3x = 4 + 4y$

25. $3p = 2q + 1$
$\quad 2p + 3q = 5$

26. $m = -3$
$\quad 4m + 7n = 2$

27. $d + 3t = 16$
$\quad 7d + 4t = 10$

Determine whether each of the following systems has a single solution, no solution, or an unlimited number of solutions:

28. $a = -b$
$\quad a + b = 0$

29. $6u - 3v = 7$
$\quad 30u - 15v = 27$

30. $y = 3x - 2$
$\quad y = 6x - 4$

31. $m - 3n = 5$
$\quad 2m - 3n = 10$

32. $5p - 3q = 7$
$\quad 65p - 51q = 119$

33. $3c - 2d = 13$
$\quad 2c - 3d = 13$

34. $8x - 3y = 11$
$\quad 8x + 3y = 11$

35. $4r + 7s = 26$
$\quad 4r + 7s = 13$

36. $y = 5x - 2$
$\quad y = 5x - 2$

37. Determine five different solutions for the system
$$2x + 3y = 24$$
$$4x + 6y = 48$$

19   Verbal Problems in Two Unknowns

When a verbal problem concerns two unknowns, it is often expedient to use a separate literal symbol to represent each unknown.   The conditions of the problem must then be interpreted to provide two equations involving the unknowns.   If the solution of these equations checks against the stated requirements, the problem is solved; if not, the values must be rejected, and no solution is possible.   The examples which follow illustrate both situations.

EXAMPLE 1.   Find two numbers such that twice the first added to the second is 30, and three times the first exceeds the second by 25.

$$\text{Let } x = \text{first number}$$
$$\text{and } y = \text{second number}$$

The equations are

$2x + y = 30$ (twice the first added to the second is 30)
$3x - y = 25$ (three times the first exceeds the second by 25)

Solving this system, we obtain

$$x = 11$$
$$y = 8$$

The two numbers 11 and 8 satisfy the conditions of the original problem since twice 11 added to 8 is 30, and three times 11 is 25 more than 8.

EXAMPLE 2.   Two tanks holds 28,000 and 35,000 gallons of oil, respectively.   Oil is being pumped from the first tank at a rate of 14 gallons per minute, and from the second at a rate of 12 gallons per minute.   How long will it be before they have equal volumes of oil remaining, and what amount of oil will remain in each tank?   Let x = number of minutes required to reach equal volumes of oil and y = number of gallons in each tank when the remaining volumes are equal.   The equations are

$28000 - 14x = y$ (for first tank, original volume minus amount pumped out in x minutes equals final volume)
$35000 - 12x = y$ (for second tank, original volume minus amount pumped out in x minutes equals final volume)

Solving these equations, we find

$$x = -3500 \text{ minutes}$$

This is the only possible value of x, but it must be rejected since the problem does not have meaning for negative values of time. Therefore the problem, as stated, has no solution.

EXAMPLE 3.   The sum of the digits of a two-digit number is 11. If the digits are reversed, the new number is 27 less than the original.   Find the number.

Let x = digit in ten's place and y = digit in unit's place.   Then

$10x + y$ = value of the original number and $10y + x$ = value of the number obtained by reversing digits.

The equations are

$$x + y = 11 \text{ (sum of the digits is 11)}$$
$$(10x + y) - (10y + x) = 27 \text{ (new number is 27 less than the original)}$$

Rewriting the second equation, the system becomes

$$x + y = 11$$
$$9x - 9y = 27$$

Solving this system leads to the values

$$x = 7$$
$$y = 4$$

Therefore the number is 74.   This number satisfies the conditions of the problem since the sum of its digits is 11 and, reversing the digits, 47 is 27 less than 74.

Exercise 19

PROBLEMS IN TWO UNKNOWNS

Solve each of the following exercises by introducing two unknowns.

1. The sum of two numbers is 210.   The larger exceeds the smaller by 44.   What are the numbers?

2. The total capacity of two storage tanks is 775 gallons.   If one tank can hold 225 gallons more than the other, find the capacity of each tank.

3. The difference of two numbers is 25, and twice the smaller is 13 less than the larger.   Find the numbers.

4. The acute angles of a right triangle total 90°.   If one of these angles is 30° greater than three times the other, find the size of each acute angle.

5. Separate 144 into two parts such that the larger is 7 times the smaller.

6. The length of a rectangle exceeds its width by 14 inches.   Find the length and width, if the perimeter is 60 inches.

7. In the equation $y = ax + b$, it is known that $y = 19$ when $x = -3$ and $y = -1$ when $x = 2$.   Find the values of a and b.

8. The sum of the present ages of a man and his son is 62 years. Five years from now, the man will be twice as old as his son.   Find their present ages.

9. A number between 10 and 100 is reduced by 54 if its digits are reversed. Find the number if the sum of its digits is 10.

10. Seven adult tickets and 6 children's tickets cost $18.05, and 5 adult tickets and 9 children's tickets cost $16.90. What is the price of each kind of ticket?

11. A collection of 25 coins totals $2.65. If the collection consists of nickels and quarters, how many of each are there?

12. The area of a trapezoid is 136 square inches. Its altitude is 8 inches. Find the lengths of the bases, if their difference is 8 inches.

13. A two-digit number is six times the sum of its digits, and if 9 is subtracted from the number, the digits will be reversed. What is the number?

14. A group of businessmen purchase a parking lot for the use of their customers. They agree to pay equal shares. If 2 more had subscribed to the project, each would have paid $100 less, but if 4 less had subscribed, each would have paid $300 more. How many men were involved in the transaction, and what was the cost of the lot?

15. A merchant has $45 in quarters and dimes. How many coins of each kind has he if the total number is 240?

16. A part of $6700 was invested at $4\frac{1}{2}\%$ and a part at 5%. The $4\frac{1}{2}\%$ investment yields $45 more per year than the 5% investment. Find the amount invested at each rate.

17. Three years ago, John was three times as old as his sister. In 2 years, he will be twice as old as his sister. Find the present age of each.

18. The admission to a community club dance was $2.50 for members and $4.00 for nonmembers. The total receipts were $1470. If there were 510 tickets sold, how many of each kind were sold?

19. The perimeter of a rectangle is 328 feet. If the length is increased by 1 foot and the width is decreased by 5 feet, the area of the rectangle will be decreased by 585 square feet. Find the dimensions of the rectangle.

20. The difference of the present ages of a father and his daughter is 24 years. Nine years ago, the father was three times as old as his daughter. What are their present ages?

21. A chemist has a 90% solution of sulphuric acid and a 70% solution of the same acid. How many gallons of each must be mixed to make 200 gallons of a 75% solution?

22. A steamer goes downstream 70 miles in 5 hours. The return trip takes 2 hours longer. What is the rate of the steamer in still water, and what is the rate of the current?

23. An investment of $12,000 yields $635 per year. If part of the $12,000 is invested at 5% and the balance at 6%, find the amount invested at each rate.

24. A plane makes a trip of 450 miles with the wind in 1.5 hours and returns against the wind in 1.8 hours. Find the speed of the plane in still air and the speed of the wind.

25. A radiator contains 20 quarts of a 30% solution of antifreeze. How many quarts must be drained off and replaced by a 90% antifreeze solution to obtain 20 quarts of a 66% solution?

26. A certain company supplies natural gas for space heating at a fixed monthly service charge plus a given rate per thousand cubic feet of gas. The Browns were billed $19.50 for the month of February when they used 225,000 cubic feet of gas and $11.90 for the month of March when they used 130,000 cubic feet. Find the amount of the service charge and the rate per thousand cubic feet.

CHAPTER **5**

Factoring and the Solution of Quadratic Equations

We have previously discussed methods of finding the product of polynomials. Certain of these products occur so frequently that it is convenient to develop more rapid methods of multiplication for these cases.

The Product of Two Binomials of the Form $(ax + by)(cx + dy)$

It will be recalled, from our consideration of the multiplication of polynomials, that the product $(ax + by)(cx + dy)$ can be determined as follows:

$$ax + by$$
$$cx + dy$$
$$\overline{acx^2 + bcxy}$$
$$\underline{\qquad + adxy + bdy^2}$$
$$acx^2 + (bc + ad)xy + bdy^2$$

Note that the middle term of the product, $(bc + ad)xy$, follows as a consequence of the distributive law, since $bcxy + adxy = (bc + ad)xy$.

We may arrive at this product more rapidly by observing the following scheme (the small numbers indicate the order of multiplication):

100

$$(ax + by)(cx + dy) = acx^2 + adxy + bcxy + bdy^2$$

$$= acx^2 + (ad + bc)xy + bdy^2$$

As an example, we have

$$(2x + 3y)(4x - 7y) = 8x^2 - 14xy + 12xy - 21y^2 = 8x^2 - 2xy - 21y^2$$

We may state this procedure in words:

In finding the product of two binomials $(ax + by)(cx + dy)$:

(*a*) The first term of the product is the product of the first term of each binomial.

(*b*) The second term is the sum of the products of the "outer" and "inner" terms of the two binomials.

(*c*) The last term is the product of the last term of each binomial. After a little practice, the addition of the two terms in step *b* should be done mentally. The following examples show that this method of multiplication applies to any two binomials.

EXAMPLE 1. $(2x + 3y)(3x + 4y) = 6x^2 + 8xy + 9xy + 12y^2$
$$= 6x + 17xy + 12y^2$$

EXAMPLE 2. $(4t - 5)(2t - 7) = 8t^2 - 38t + 35$

EXAMPLE 3. $(2 + 7x)(3 - 4x) = 6 + 13x - 28x^2$

The Square of a Binomial: $(ax + by)^2$

Although the preceding method of multiplication applies to any two binomials, we consider the special cases

$$(ax + by)^2 = (ax + by)(ax + by) = a^2x^2 + 2abxy + b^2y^2$$
$$(ax - by)^2 = (ax - by)(ax - by) = a^2x^2 - 2abxy + b^2y^2$$

We may state this procedure in words:

> The square of a binomial equals the square of the first term plus twice the product of the two terms plus the square of the last term.

We note that the middle term will be positive if the binomial is the sum of two terms; the middle term will be negative if the binomial is the difference of two terms.

EXAMPLE 4. $(2x + 3y)^2 = 4x^2 + 12xy + 9y^2$

EXAMPLE 5. $(2x - 3y)^2 = 4x^2 - 12xy + 9y^2$

EXAMPLE 6. $(3x - 7)^2 = 9x^2 - 42x + 49$

The Product of the Sum and Difference of Two Terms: $(ax + by)(ax - by)$

The product $(ax + by)(ax - by)$ is another special case of the product of two binomials.   Since

$$(ax + by)(ax - by) = a^2x^2 - abxy + abxy - b^2y^2 = a^2x^2 - b^2y^2$$

we have the following rule:

> The product of the sum and difference of two terms equals the square of the first term minus the square of the last term.

EXAMPLE 7. $(3x + 7y)(3x - 7y) = 9x^2 - 49y^2$

EXAMPLE 8. $(2t - 3)(2t + 3) = 4t^2 - 9$

EXAMPLE 9. $(3 - 5w)(3 + 5w) = 9 - 25w^2$

Application of Special Products to Arithmetic Computation

If we wish to find the square of 23, we may write

$$(23)^2 = (20 + 3)^2 = 400 + 120 + 9 = 529$$

To find the square of 29, we may write

$$(29)^2 = (30 - 1)^2 = 900 - 60 + 1 = 841$$

Certain products may be easily found, as, for example,

$$(23)(17) = (20 + 3)(20 - 3) = 400 - 9 = 391$$

Exercise 20

SPECIAL PRODUCTS

Perform the following multiplications:

1. $(2x + 3y)(3x + 4y)$

2. $(3x - 7y)(2x - 5y)$

3. $(3x - 2y)(2x - 3y)$

4. $(4x - 7)(2x - 3)$

5. $(2t - 3)(2t + 7)$

6. $(2t - 3)(2t - 7)$

7. $(x + 7)(x + 3)$

8. $(x - 7)(x - 3)$

9. $(x + 7)(x - 3)$

10. $(x - 7)(x + 3)$

11. $(a + b)(a + 2b)$

12. $(x + 2t)(x - 4t)$

13. $(3x - 2y)(x - y)$

14. $(3x - 2y)(x + y)$

15. $(3x + 2y)(x + y)$

Find, as indicated, the square of each of the following binomials:

16. $(2x - 3y)^2$ **17.** $(3x - 4)^2$ **18.** $(4 - 3x)^2$ **19.** $(2t - 7)^2$

20. $(2t + 7)^2$ **21.** $(4 + 3x)^2$ **22.** $(\pi - h)^2$ **23.** $(\pi + h)^2$

24. $(2\pi r + h)^2$ **25.** $(r + r')^2$ **26.** $(r - r')^2$ **27.** $(b_1 + b_2)^2$

Perform the following multiplications:

28. $(3x - 2y)(3x + 2y)$

29. $(2x - 7)(2x + 7)$

30. $(3t + 4)(3t - 4)$

31. $(t^2 - 6)(t^2 + 6)$

32. $(\pi r + h)(\pi r - h)$

33. $(r + r')(r - r')$

34. $(y - 2)(y + 2)$

35. $(y^2 - 3)(y^2 + 3)$

36. $(x^2 - y^2)(x^2 + y^2)$

Find each of the following squares by expressing the base as a binomial:

37. $(35)^2$ **38.** $(27)^2$ **39.** $(21)^2$ **40.** $(39)^2$ **41.** $(16)^2$

42. $(57)^2$ **43.** $(63)^2$ **44.** $(89)^2$ **45.** $(401)^2$ **46.** $(399)^2$

47. $(798)^2$ **48.** $(191)^2$

Find each of the following products by writing the factors as the sum and difference of two terms:

49. $(24)(16)$ **50.** $(25)(35)$ **51.** $(29)(31)$ **52.** $(62)(58)$

53. $(21)(19)$ **54.** $(82)(78)$ **55.** $(38)(42)$ **56.** $(210)(190)$

Determine the missing term which will make each of the following expressions the square of a binomial:

57. $x^2 - (?) + 81y^2$ **58.** $x^2 - 14xy + (?)$ **59.** $9x^2 - 42x + (?)$

60. $(?) + 70x + 25$ **61.** $4t^2 - (?) + 1$ **62.** $16t^2 - 8t + (?)$

63. $16m^2 - (?) + 25$ **64.** $y^2 + (?) + \frac{1}{4}$ **65.** $(?) + 42xy + 49y^2$

Find each of the following products:

66. $(x + y + 7)(x + y + 9) = [(x + y) + 7][(x + y) + 9] = (x + y)^2$
$+ 16(x + y) + 63 = x^2 + 2xy + y^2 + 16x + 16y + 63.$

67. $(x + y - 2)(x + y + 2)$ **68.** $(x + y + 3)(x + y + 3)$

69. $(x + y + 2)(x + y + 3)$

70. $(x + y + 1)(x - y - 1) = [x + (y + 1)][x - (y + 1)] = x^2 -$
$(y + 1)^2$ etc.

71. $(2s + t + 3)(2s - t - 3)$ **72.** $(4t + n - 6)(4t - n + 6)$

73. $(3x + 2y - 1)(3x - 2y + 1)$ **74.** $(1 + s + t)(1 - s - t)$

Find each of the following products:

75. $(n - 2)(n^2 + 4)(n + 2)$ **76.** $(x + 1)(x - 1)(x^2 + 1)$

77. $(n + 2)(n - 2)(n^2 - 4)$ **78.** $(x + 1)(x - 1)(x^2 - 1)$

21 Factoring

Factoring any expression, whether it is a number such as 1073 or a trinomial such as $3x^2 + 17xy + 10y^2$, is the process of finding two or more numbers (or algebraic expressions) whose product equals the given expression. The factors of 1073 are 29 and 37, since $(29)(37) = 1073$, and the factors of $3x^2 + 17xy + 10y^2$ are $3x + 2y$ and $x + 5y$, since $(3x + 2y)(x + 5y) = 3x^2 + 17xy + 10y^2$. The process of factoring, closely related to the process of multiplication, is more difficult than multiplication since in most cases we must resort to trial and error. When we multiply 29 by 37, we have a definite process whereby we arrive at the product 1073. The converse problem of finding two (or more) numbers whose product is 1073 has no similar process. The fact that we cannot be certain that we have correctly factored an expression until we have verified our result by multiplication is one of the reasons for our study of the special products in the previous section. In our discussion of methods of factoring, we shall limit ourselves to certain expressions which occur most frequently in algebra.

Factoring the Trinomial: $ax^2 + bxy + cy^2$

Let us again consider the factors of $3x^2 + 17xy + 10y^2$. If the factors are to be of the form $(ax + by)(cx + dy)$, the product of the first terms must equal $3x^2$, and the product of the last terms must equal $10y^2$, in accordance with the rule for special products of this type. The possible factors are then:

$$(a) \quad (3x + y)(x + 10y) = 3x^2 + 31xy + 10y^2$$
$$(b) \quad (3x + 10y)(x + y) = 3x^2 + 13xy + 10y^2$$
$$(c) \quad (3x + 2y)(x + 5y) = 3x^2 + 17xy + 10y^2$$
$$(d) \quad (3x + 5y)(x + 2y) = 3x^2 + 11xy + 10y^2$$

Note that only the third product, $(3x + 2y)(x + 5y)$, gives the proper second term, $17xy$. It is clear, therefore, that in factoring expressions of the type $ax^2 + bxy + cy^2$ (or of the type $ax^2 + bx + c$), we must apply the following rule:

1. The product of the first term of each binomial factor must equal the first term of the trinomial.

2. The product of the last term of each binomial factor must equal the last term of the trinomial.

3. The sum of the products of the "outer" and "inner" terms of the two factors must equal the second term of the trinomial.

EXAMPLE 1. Factor $3x^2 - 11x + 6$. The factors of the first term are $3x$ and x. The factors of the last term are 6 and 1, -6 and -1, 3 and 2, and -3 and -2. Since the first and last terms are positive, and the middle term is negative, that is, $-11x$, we need only try the negative factors of 6. The possible factors are then:

$$(a) \quad (3x - 1)(x - 6) = 3x^2 - 19x + 6$$
$$(b) \quad (3x - 6)(x - 1) = 3x^2 - 9x + 6$$
$$(c) \quad (3x - 2)(x - 3) = 3x^2 - 11x + 6$$
$$(d) \quad (3x - 3)(x - 2) = 3x^2 - 9x + 6$$

We see that only the third product has the proper second term, $-11x$, so that the factors of $3x^2 - 11x + 6$ are $(3x - 2)(x - 3)$. One more remark may be appropriate here. We note that the first factor in b and in d is divisible by 3. Since $3x^2 - 11x + 6$ is not divisible by 3, we could have immediately ruled out b and d as possible factors.

EXAMPLE 2. Factor $18x^2 + 45x - 50$. The possible combinations which will give $18x^2$ for the first term and -50 for the last term are numerous. The only combination, however, which gives $+45x$ for the second term is $(6x - 5)(3x + 10)$, which are the factors of the given expression.

Factoring the Trinomial Square: $a^2x^2 + 2abxy + b^2y^2$

Let us consider the factors of $16x^2 + 24xy + 9y^2$. Since $16x^2$ and $9y^2$ are perfect squares, we try as our first combination the factors $(4x + 3y)(4x + 3y)$. Since the product of these factors equals $16x^2 + 24xy + 9y^2$, we have

$$16x^2 + 24xy + 9y^2 = (4x + 3y)(4x + 3y) \quad \text{or} \quad (4x + 3y)^2$$

We note that a trinomial whose first and last terms are perfect squares need not be itself a perfect square. Thus

$$36x^2 - 97x + 36 = (9x - 4)(4x - 9); \text{ not } (6x - 6)(6x - 6)$$

which emphasizes the fact that every factoring problem must be checked by multiplication of the factors. From the rule for the square of a binomial, however, it is not difficult to determine in advance if a trinomial is a perfect square. The student will be asked to formulate a rule for doing this in Problem 66 of Exercise 21.

EXAMPLE 3. $4x^2 - 12x + 9 = (2x - 3)(2x - 3) = (2x - 3)^2$

EXAMPLE 4. $144 - 72x + 9x^2 = (12 - 3x)(12 - 3x) = (12 - 3x)^2$

Factoring the Difference of Two Squares: $a^2x^2 - b^2y^2$

Let us consider the factors of $9x^2 - 49y^2$. As a direct consequence of the rule for the product of the sum and difference of two terms, we may write at once

$$9x^2 - 49y^2 = (3x + 7y)(3x - 7y)$$

The student will be asked to formulate a rule for factoring the difference of two squares in Problem 76 of Exercise 21.

EXAMPLE 5. $16x^2 - 81 = (4x - 9)(4x + 9)$

Factoring a Polynomial with a Common Monomial Factor: $ax + bx - cx$

As a direct consequence of the distributive postulate, we have

$$ax + bx - cx = x(a + b - c)$$

EXAMPLE 6. $3x^3 - 6x^2 + 15x = 3x(x^2 - 2x + 5)$

EXAMPLE 7. $\pi r^2 + 2\pi rh = \pi r(r + 2h)$

The student will be asked to formulate a rule for factoring an expression with a common monomial factor in Problem 77 of Exercise 21.

Exercise 21

FACTORING

Factor each of the following trinomials:

1. $x^2 + 3x + 2$
2. $x^2 - 4x + 3$
3. $2n^2 + 7n + 6$
4. $2y^2 + 9y + 9$
5. $3x^2 + 8x + 4$
6. $a^2 - a - 12$
7. $x^2 + 5xy - 24y^2$
8. $x^2 - 9xy - 22y^2$
9. $x^2 - 3x - 28$
10. $2x^2 + 5xy + 3y^2$
11. $2x^2 + 7xz + 3z^2$
12. $2l^2 - 5ly - 3y^2$
13. $2x^2 - 3x - 2$
14. $2x^2 - 3x - 9$
15. $8x^2 + 10xy - 3y^2$
16. $6x^2 + 13x - 5$
17. $7x^2 - 41x - 6$
18. $3a^2x^2 + 2ax - 1$
19. $2x^2 - x - 6$
20. $x^2 + 42x - 43$
21. $x^2 - 42x - 43$
22. $x^2 - 4xy + 4y^2$
23. $l^2 - l - 30$
24. $2x^2 + 5xy + 2y^2$

Factor each of the following trinomial squares:

25. $a^2 + 4a + 4$
26. $x^2 + 22x + 121$
27. $16l^2 + 8l + 1$
28. $x^2 - 10xy + 25y^2$
29. $16l^2 - 8al + a^2$
30. $4x^2 - 12xy + 9y^2$
31. $25x^2 - 20x + 4$
32. $9x^2 + 30tx + 25t^2$
33. $169x^2 + 78x + 9$
34. $9y^2 + 66y + 121$
35. $9 - 6a + a^2$
36. $m^2n^2 - 16mn + 64$
37. $x^2 - 20x + 100$
38. $x^2 + 36x + 324$
39. $4m^2 + 4m + 1$

Factor each of the following binomials, which are the differences of two squares:

40. $x^2 - y^2$
41. $a^2 - 1$
42. $4l^2 - 9$
43. $4l^2 - 25r^2$
44. $144x^2 - 49y^2$
45. $36c^2d^2 - 9$
46. $x^2 - \frac{1}{9}$
47. $2.25a^2 - 1.69b^2$
48. $0.81x^2 - 1.96y^2$

Factor each of the following expressions by removing the common monomial factor:

49. $a^3 + a^2$
50. $3l + 6$
51. $3x - 6$
52. $ax + a$
53. $3ax - 6a$
54. $2\pi R - 2\pi r$
55. $2\pi RH + 2\pi rh$
56. $\pi R^2 - \pi r^2$
57. $\pi r^2 + 2\pi r$
58. $am + an + ax$
59. $3c^2 - 12c - 18c^4$
60. $ax - a^2x^2 - a^3x$
61. $5c^2 + 10c - 15$
62. $5x - 10x^2 + 15x^3$
63. $6x^3y + 3xy^2 - 9x^2y^2$
64. $\pi R^2h + \pi r^2h + \pi Rrh$
65. $2a^2 + 2a^3 + 2a^4$
66. State a rule for determining if a trinomial is a perfect square.

Determine, without factoring, which of the following trinomials are perfect squares:

67. $x^2 - 4x + 9$ **68.** $x^2 - 4x + 4$
69. $4x^2 - 12xy + 9y^2$ **70.** $4x^2 + 12xy + 9y^2$
71. $9x^2 - 14x + 16$ **72.** $9x^2 - 24xy + 16y^2$
73. $a^2x^2 + 2axy + y^2$ **74.** $25x^2 - 50x + 100$
75. $25x^2 - 100xy + 100y^2$

76. State a rule for factoring the difference of two squares.

77. State a rule for factoring an expression with a common monomial factor.

Factor each of the following expressions:

78. $(x + y)^2 - 1 = (x + y + 1)(x + y - 1)$
79. $(a + t)^2 - 4$ **80.** $(2x + 3)^2 - 9y^2$ **81.** $4(x - 3)^2 - 25$
82. $16x^2 - 9(y + 1)^2$ **83.** $x^2 + 2x + 1 - t^2$ **84.** $4t^2 - 12t + 9 - r^2$
85. $x^2 - a^2 + 2a - 1$

22 Prime Factors

Prime Factors of an Algebraic Expression

An algebraic expression is called *prime* if it has no other factors than itself (or its negative) and 1.   Examples of prime expressions are $3ax + b$, $x^2 + 7$, and $x^2 - 3x + 7$.   We must observe that although $x^2 - 7 = (x + \sqrt{7})(x - \sqrt{7})$, we do not consider such factors in this discussion, and therefore $x^2 - 7$ is prime.   An expression which is not prime is called *composite:* $x^2 - 25$ and $x^2 - 3x + 2$ are composite expressions.

Let us consider the factors of $2x^2 - 6x + 4$.   Removing the common monomial factor 2, we have

$$2x^2 - 6x + 4 = 2(x^2 - 3x + 2)$$

But the second factor is composite, and its factors are $(x - 2)(x - 1)$. Hence

$$2x^2 - 6x + 4 = 2(x - 2)(x - 1)$$

We say that 2, $x - 2$, and $x - 1$ are the prime factors of $2x^2 - 6x + 4$.

Suppose we had first written the factors as

$$2x^2 - 6x + 4 = (2x - 4)(x - 1)$$

We note that 2 is a common monomial factor of $2x - 4$, so that

$$2x^2 - 6x + 4 = 2(x - 2)(x - 1)$$

as before.   This illustrates an important theorem of algebra:

A polynomial has one and only one set of prime factors.*

Whenever an expression has a common monomial factor, it is always best to remove this factor before any further factoring is tried.

EXAMPLE 1.

$$18x^2 - 98y^2 = 2(9x^2 - 49y^2) = 2(3x - 7y)(3x + 7y)$$

EXAMPLE 2.

$$18x^2 - 84xy + 98y^2 = 2(9x^2 - 42xy + 49y^2) = 2(3x - 7y)^2$$

EXAMPLE 3.

$$16x^4 - y^4 = (4x^2 - y^2)(4x^2 + y^2) = (2x - y)(2x + y)(4x^2 + y^2)$$

EXAMPLE 4. $a^4 - 13a^2 + 36 = (a^2 - 9)(a^2 - 4)$
$$= (a - 3)(a + 3)(a - 2)(a + 2)$$

Exercise 22

PRIME FACTORS

Express each of the following expressions as the product of its prime factors:

1. $3x^2 - 12x + 12$
2. $2x^2 + 4x - 6$
3. $4a^2 - 144$
4. $3ax^2 - ax - 2a$
5. $6x^2 + 34x + 20$
6. $6x^2 - 6x - 72$
7. $2x^2 - 8y^2$
8. $x^5 + 12x^3 + 36x$
9. $2n^3 - 2n$
10. $ax^4 - ax^8$
11. $2x^8 - 2$
12. $3x^2 - 27y^2$
13. $4n^4 - 9n^2 - 9$
14. $4a^4 - 92a^2 - 200$
15. $5an^4 - 20n^5$
16. $2x^2y - 6xy^2 - 140y^3$
17. $x^4 - 4x^2 + 4$
18. $8a^2 - 30a - 8$
19. $5x^3 - 20xy^2$
20. $x^4 - 6x^2 - 27$
21. $2ax^2 + 28ax + 98a$
22. $4x^4 - 13x^2 + 9$
23. $24x^2 + 6xy - 18y^2$
24. $2a + ax - 3ax^2$
25. $5x^4 - 80$
26. $2x^2 - 4x - 126$
27. $8x^4 - 26x^2 + 18$
28. $3m^2 - 30m + 63$
29. $m^3 - m^2 - 30m$
30. $5x^3 - 20x^2 - 300x$

* H. B. Fine, *College Algebra*, Dover, 1905, page 210.

1. Look for common monomial factor.
2. Apply factoring forms.

Express each of the following formulas in factored form:

31. $P = 2l + 2w$ | Perimeter of a rectangle.
32. $A = \pi R^2 - \pi r^2$ | Area of a washer.
33. $S = \pi r^2 + \pi rs$ | Total surface of a right circular cone.
34. $S = 2\pi r^2 + 2\pi rh$ | Total surface of a right circular cylinder.
35. $S = n^2 + n$ | Sum of the first "n" even numbers.
36. $a^2 = c^2 - b^2$ | Pythagorean Theorem.
37. $A = \frac{1}{2}hb + \frac{1}{2}hb'$ | Area of a trapezoid.
38. $C = \frac{5}{9}F - \frac{160}{9}$ | Centigrade temperature in terms of Fahrenheit temperature.
39. $A = P + Prt$ | Amount of P dollars at r per cent interest for t years.

40. Given a washer with outer radius equal to $2\frac{1}{2}$ inches and inner radius equal to $1\frac{1}{2}$ inches. Using $\pi = 3.14$ find the area by using the formula of Problem 32 in both unfactored and factored form. Which form involves the least amount of computation?

Factor each of the following expressions:

41. $a^2 + 2ab + b^2 - c^2 = (a + b)^2 - c^2 = (a + b + c)(a + b - c)$
42. $4a^2 + 4a + 1 - b^2$ **43.** $x^2 - y^2 + 2y - 1$
44. $x^2 + y^2 - z^2 - 2xy$ **45.** $x^2 + y^2 - z^2 + 2xy$
46. $x^2 - 4a^2 - 4a - 1$ **47.** $a^2 + 2ab + b^2 - m^2 - 2mn - n^2$
48. $a^2 + 2ab + b^2 - m^2 + 2mn - n^2$
49. $4x^2 + 4x + 1 - y^2 - 2y - 1$
50. $4 - 9x^2 + 54xy - 81y^2$

23   Quadratic Equations

Definition. A quadratic equation in one variable is an equation in which the highest power of the variable is two. Examples of quadratic equations are:

$$(1) \quad x^2 - 3x + 5 = 0$$
$$(2) \quad 3x^2 - 2x = 7$$
$$(3) \quad 3x^2 - 6 = 0$$
$$(4) \quad 2x^2 = 8x$$
$$(5) \quad 4x^2 = 0$$
$$(6) \quad kx^2 + (m + n)x + p = 0$$

(handwritten annotations at top: $ax^2 + bx + c = 0$ with "quadratic term", "constant", "linear term" labels; "incomplete quadratic if 2 or 3 term missing")

Standard Form of a Quadratic Equation

Although it is not necessary for the right-hand member of a quadratic equation to equal zero, it is convenient to write the equation in this way, and we represent the general quadratic equation in one variable as

$$ax^2 + bx + c = 0 \qquad a \neq 0$$

We observe that a is the coefficient of x^2, b is the coefficient of x, and c is the constant term. The coefficient a cannot equal zero since the equation would then be a linear equation. The coefficient b and the constant term c may assume any value, including zero. If b is zero the resulting equation is called a pure quadratic equation.

If the equation is not written in standard form, we can always rewrite it in this form. When we do so, it is customary to have a positive, as in the following examples:

	Equation	Standard Form
(1)	$2x^2 - 3x = 5$	$2x^2 - 3x - 5 = 0$
(2)	$6x - 7 = 3x^2$	$3x^2 - 6x + 7 = 0$
(3)	$-2x^2 + 7x = 8$	$2x^2 - 7x + 8 = 0$
(4)	$3x^2 = 6x$	$3x^2 - 6x = 0$
(5)	$4x^2 = 9$	$4x^2 - 9 = 0$

Although a, b, and c need not be integers, we may always write an equivalent equation in which the coefficients are integers by multiplying the equation by a suitable number. Thus

$$\tfrac{2}{3}x^2 + 7x - 4 = 0 \tag{1}$$

is equivalent to

$$2x^2 + 21x - 12 = 0 \tag{2}$$

since each member of Equation 1 is multiplied by 3. We may also write any quadratic equation so that $a = 1$ by dividing both members of the equation by a. Thus

$$3x^2 - 7x + 7 = 0 \tag{3}$$

is equivalent to

$$x^2 - \tfrac{7}{3}x + \tfrac{7}{3} = 0 \tag{4}$$

since each member of Equation 3 is divided by 3.

Roots of the Quadratic Equation

The values of x which satisfy the quadratic equation $ax^2 + bx + c = 0$ are called the roots of the equation. Every quadratic equation has two roots; that is, there are two values of x which satisfy the equation. For example, the equation

$$x^2 - 3x + 2 = 0$$

is satisfied by $x = 1$ and $x = 2$, as may be readily verified by substituting 1 and 2 for x.

The two roots of the equation, however, need not be distinct; that is, they may equal each other. For example, the equation

$$x^2 - 2x + 1 = 0$$

is satisfied for $x = 1$, and for no other value of x. We say that the equation has a double root, namely, $x = 1$ and $x = 1$.

Exercise 23

QUADRATIC EQUATIONS

Which of the following equations are quadratic equations? (Letters other than x or t are to be considered here as constants.)

1. $x^2 - 3x + 7 = 0$	**2.** $9x - 7 = 0$	**3.** $a^2x + b = c$
4. $a^2x^2 + bx + c = 0$	**5.** $6x^2 - 7 = 0$	**6.** $ax^2 + b + c = 0$
7. $t^2 + 2t - 6 = 0$	**8.** $t^2 = 6t - 7$	**9.** $t^2 + 17t = 16t$

Rewrite each of the following equations in standard form:

10. $3x^2 - 7x = 7$	**11.** $2x^2 = 6 - 3x$	**12.** $6x - x^2 = 17$
13. $3x = x^2 - 17$	**14.** $6x^2 = 7$	**15.** $x^2 = 19 + 2x$
16. $6x = 13 + 2x^2$	**17.** $6x = 13 - 2x^2$	**18.** $x - 3x^2 = 12$
19. $17 = 12x + x^2$	**20.** $17 = 12x - x^2$	**21.** $17x = 12 - x^2$
22. $x^2 - 6 = 6x$	**23.** $x^2 + 6 = 6x$	**24.** $(x - 2)^2 - x = 1$

State the values of a, b, and c, in each of the following equations:

25. $2x^2 - 7x + 7 = 0$	**26.** $x^2 - 17\dot{x} - 7 = 0$	**27.** $x^2 + 3x - 4 = 0$
28. $2x^2 = 7x - 3$	**29.** $3x^2 - 4x = 4$	**30.** $2x - 4 = x^2$
31. $3x = 7 - 4x^2$	**32.** $5x^2 - 5x - 5 = 0$	**33.** $ax^2 - dx + k = 0$

Determine by substitution which of the numbers to the right of each equation is a root of the equation:

34. $x^2 + 3x + 2 = 0$ $(1, -1, 2, -2)$

35. $x^2 - 4x - 5 = 0$ $(1, 2, -1, -2, -5)$

36. $2x^2 - 8 = 0$ $(2, -2, 4, -4)$

37. $3x^2 - 75 = 0$ $(1, -1, 3, -3, 5, -5)$

38. $3x^2 + 13x = 0$ $(0, 3, -3, \frac{13}{3}, -\frac{13}{3})$

39. $x^2 + x - 6 = 0$ $(1, -1, 2, -2, 3, -3)$

40. $3x^2 - 7x + 2 = 0$ $(1, -1, 2, -2, \frac{1}{3}, -\frac{1}{3})$

41. $7x^2 - 23x + 6 = 0$ $(1, -1, 3, -3, \frac{2}{7}, -\frac{2}{7})$

42. $4x^2 - 12x + 9 = 0$ $(\frac{2}{3}, -\frac{2}{3}, \frac{3}{2}, -\frac{3}{2})$

43. $x^2 + x + 5 = 0$ $(1, -1, 5, -5)$

Find by trial the roots of the following pure quadratic equations:

44. $x^2 = 25$ **45.** $3x^2 - 27 = 0$ **46.** $x^2 = \frac{4}{9}$

47. $6x^2 - 54 = 0$ **48.** $4x^2 = 9$ **49.** $8x^2 - 32 = 0$

50. After observing the results of Problems 44 to 49, what may be said about the roots of the quadratic equation $ax^2 + bx + c = 0$ when $b = 0$ and c is negative?

Determine by substitution which of the following equations have zero as one root:

51. $9x^2 - 3x = 0$ **52.** $9x^2 - 2x + 7 = 0$ **53.** $6x^2 = 7x$

54. $6 = x - 2x^2$ **55.** $0 = 6x - 2x^2$ **56.** $9x^2 = 9$

57. After observing the results of Problems 51 to 56, what must be the value of c in the equation $ax^2 + bx + c = 0$ when one of the roots equals zero?

Rewrite each of the following equations in standard form with all of the coefficients integers:

58. $3x^2 - \frac{2}{3}x + 8 = 0$ **59.** $2x^2 - 3x = \frac{1}{3}$ **60.** $\frac{2}{3}x^2 - 3x + 1 = 0$

61. $x^2 - \frac{1}{5}x = \frac{1}{5}$ **62.** $\frac{1}{4}x^2 - \frac{1}{2}x = 2$ **63.** $\frac{1}{3}x^2 - \frac{2}{3}x = 5$

Rewrite each of the following equations so that a equals 1:

64. $3x^2 - 2x + 5 = 0$ **65.** $2x^2 - 4x + 7 = 0$

66. $\frac{1}{3}x^2 - 2x + 4 = 0$ **67.** $5x^2 - 3x - 12 = 0$

68. $0.2x^2 - 0.3x = 1.4$ **69.** $0.04x^2 - 0.02x = 1$

24   Solution of the Quadratic Equation by Factoring

In Chapter 4, the solution of linear equations was effected by systematic application of the four fundamental operations and the postulates of equality.   We shall now show that the solution of quadratic equations may be made to depend on the solution of related linear equations.   Let us consider the equation

$$x^2 - 3x + 2 = 0 \tag{1}$$

Factoring the left-hand member, we have

$$(x - 1)(x - 2) = 0 \tag{2}$$

We now apply the following postulate:

Postulate 12.   *The product of two or more factors equals zero if and only if one or more of the factors equal zero.*

When we have done this we have

$$x - 1 = 0 \quad \text{or} \quad x - 2 = 0 \tag{3}$$

The Equations 3 are linear equations, and their solutions are

$$x = 1, \quad x = 2 \tag{4}$$

If these values of x are substituted in Equation 1, it will be found that both values satisfy the equation, so that $x = 1$ and $x = 2$ are the roots of the given equation.

Since an equation such as

$$x^2 + 2x - 5 = 0 \tag{5}$$

is not factorable in the sense discussed in Section 21, we shall not solve such equations by factoring, although it is possible to do so.*

The preceding discussion may be summarized by the following procedure for solving factorable quadratic equations:

1. Write the equation in standard form.
2. Factor the left-hand member of the equation.
3. Applying Postulate 12, set each factor equal to zero and solve the resulting linear equations.

---

* Equation 5 may be factored as $(x + 1 - \sqrt{6})(x + 1 + \sqrt{6}) = 0$, so that the roots are $x = -1 + \sqrt{6}$ and $-1 - \sqrt{6}$.

4. Check the solution by substituting the roots thus found in the original equation.

EXAMPLE 1.   Solve the equation $2x^2 - 7x + 6 = 0$.
  (1) $2x^2 - 7x + 6 = 0$
  (2) $(2x - 3)(x - 2) = 0$
  (3) $2x - 3 = 0$, or $x - 2 = 0$
        $x = \frac{3}{2}, x = 2$
  (4)   Substituting $x = \frac{3}{2}$ in the given equation,
        $2(\frac{3}{2})^2 - 7(\frac{3}{2}) + 6 = 0$
        $\frac{9}{2} - \frac{21}{2} + \frac{12}{2} = 0$
        $\qquad\qquad 0 = 0$
        Substituting $x = 2$ in the given equation,
        $2(2)^2 - 7(2) + 6 = 0$
        $8 - 14 + 6 = 0$
        $\qquad\quad 0 = 0$

EXAMPLE 2.   Solve the equation $4x^2 = 9x$.
  (1) $4x^2 - 9x = 0$
  (2) $x(4x - 9) = 0$
  (3) $x = 0$, or $4x - 9 = 0$
        $x = 0, x = \frac{9}{4}$
  (4)   Substituting $x = 0$ in the given equation,
        $4(0)^2 = 9(0)$
        $\qquad 0 = 0$
        Substituting $x = \frac{9}{4}$ in the given equation,
        $4(\frac{9}{4})^2 = 9(\frac{9}{4})$
        $\qquad \frac{81}{4} = \frac{81}{4}$

EXAMPLE 3.   Solve the equation $4x^2 - 100 = 0$.
  (1) $4x^2 - 100 = 0$
  (2) $4(x - 5)(x + 5) = 0$
  (3) $x - 5 = 0$, or $x + 5 = 0$
        $x = 5, x = -5.$
  (4)   Substituting $x = 5$ in the given equation,
        $4(5)^2 = 100$
        $100 = 100$
        Substituting $x = -5$ in the given equation,
        $4(-5)^2 = 100$
        $100 = 100$

1. standard form
2. factor out common expression
3. factor remaining
4. set it equal to zero
5. find solution set

EXAMPLE 4.   Solve the equation $4x^2 = 4x - 1$.

(1) $4x^2 - 4x + 1 = 0$

(2) $(2x - 1)(2x - 1) = 0$

(3) $2x - 1 = 0, 2x - 1 = 0$

$x = \frac{1}{2}$ (Both roots are equal to $\frac{1}{2}$)

(4)   The student should verify that $\frac{1}{2}$ is a root of the equation.

In Example 2, the student may be tempted to divide both sides of the equation by x, the resulting equation then being $4x = 9$. The root of this equation is $x = \frac{9}{4}$, as before, but the other root, $x = 0$, is lost. Thus Postulate 5 (of Section 14) for the solution of equations does not include division by expressions which contain the variable.

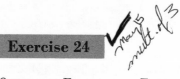

Exercise 24

SOLUTION OF THE QUADRATIC EQUATION BY FACTORING

Solve each of the following quadratic equations by factoring:

1. $x^2 = 9$ 2. $x^2 = 49$ 3. $x^2 = 7x$
4. $x^2 - 8x = 0$ 5. $x^2 - x - 42 = 0$ 6. $x^2 - x = 90$
7. $x^2 = 5x - 4$ 8. $x^2 = 5x$ 9. $x^2 = 13x + 30$
10. $x^2 - 4x + 4 = 0$ 11. $x^2 - 8x + 16 = 0$ 12. $x^2 = 6x - 9$
13. $x^2 = 3x - 2$ 14. $2x^2 - 7x = 15$ 15. $2x^2 + x = 15$
16. $3x^2 + 7x - 20 = 0$ 17. $6x^2 = 19x + 36$
18. $5x^2 + 14x - 3 = 0$ 19. $6x^2 - 13x + 5 = 0$
20. $8x^2 + 6x - 9 = 0$ 21. $4x^2 + 5x - 6 = 0$
22. $x^2 - 4ax + 4a^2 = 0$ 23. $2x^2 - ax - 6a^2 = 0$
24. $2x^2 - 3ax + a^2 = 0$

Solve each of the following cubic equations by applying Postulate 1 of this section to the three factors of each equation:

25. $x^3 - 3x^2 + 2x = 0$ 26. $x^3 - 6x^2 = -9x$
27. $x^3 = 16x$ 28. $3x^3 - 7x^2 = 20x$
29. $6x^3 - 13x^2 + 6x = 0$ 30. $2x^3 - 7x^2 = -6x$
31. $x^3 + 6x = 5x^2$ 32. $t^3 = 3t^2 - 2t$
33. $2t^3 - 7t^2 + 6t = 0$

Find the quadratic equation whose roots are the following numbers:

34. -2 and 5. *Hint:* What factors correspond to the roots $x = -2$ and $x = +5$? What is the product of these factors?

35. 3 and 2 **36.** -2 and -3 **37.** -3 and $+3$ **38.** 0 and 2

39. 0 and $\frac{1}{2}$ **40.** $\frac{2}{3}$ and $\frac{1}{2}$ **41.** $-\frac{2}{3}$ and $\frac{1}{2}$ **42.** 6 and -6

43. $\frac{2}{3}$ and $-\frac{2}{3}$ **44.** 2 and 2 **45.** -2 and -2 **46.** $\frac{1}{2}$ and $-\frac{1}{2}$

The quadratic equation $16t^2 - v_0 t - h = 0$ gives the time in seconds for a ball to reach the ground if it is thrown with an initial velocity of v_0 feet per second from a height of h feet above the ground. If the ball is thrown up v_0 is positive, if thrown down it is negative. Find the time to reach the ground for the following values of v_0 and h. Reject any negative roots. (Why?)

47. $v_0 = 210, h = 81$ **48.** $v_0 = 20, h = 176$ **49.** $v_0 = -20, h = 500$

50. $v_0 = 10, h = 0$

51. The area bounded by two concentric circles of radii R and r is given by the formula $A = \pi(R^2 - r^2)$. Find r if $A = 44\pi$ square feet and R is 12 feet.

25 Operations with Radicals

The Set of Irrational Numbers

As we have seen, a quadratic equation such as $x^2 - 2 = 0$ cannot be expressed as the product of factors containing only rational numbers, and in this sense is considered a prime expression. However, we may write

$$x^2 - 2 = (x + \sqrt{2})(x - \sqrt{2})$$

so that the roots of the equation are $x = \sqrt{2}$ and $x = -\sqrt{2}$. Since $\sqrt{2}$ cannot be expressed in the form p/q where p and q are integers,* it does not belong to the class of rational numbers. We call $\sqrt{2}$ an *irrational* number.

Definition. An irrational number is a number which cannot be expressed in the form p/q, where p and q are integers.

---

* For an interesting proof of this, see *Introduction to College Mathematics*, C. V. Newsom and H. Eves, Prentice-Hall, 1955, pages 20–21.

EXAMPLE 1. $\sqrt{3}$ is irrational.

EXAMPLE 2. $\sqrt[3]{7}$ is irrational.

EXAMPLE 3. π is irrational.

Properties of Radicals

We now state, without proof, two properties of radicals.

Property 1. $\sqrt[n]{a} \cdot \sqrt[n]{b} = \sqrt[n]{ab}$ ($a > 0$, $b > 0$, when n is even)

EXAMPLE 4. $\sqrt{20} \cdot \sqrt{5} = \sqrt{100} = 10$

EXAMPLE 5. $\sqrt[3]{4} \cdot \sqrt[3]{7} = \sqrt[3]{28}$

EXAMPLE 6. $\sqrt{27} = \sqrt{9} \cdot \sqrt{3} = 3\sqrt{3}$

We note from the examples that the equality of Property 1 may be read from left to right or from right to left.   Of particular importance is the relation

$$\sqrt{a} \cdot \sqrt{a} = \sqrt{a^2} = a$$

Property 2. $\dfrac{\sqrt[n]{a}}{\sqrt[n]{b}} = \sqrt[n]{a/b}$

EXAMPLE 7. $\dfrac{\sqrt{20}}{\sqrt{5}} = \sqrt{4} = 2$

EXAMPLE 8. $\dfrac{\sqrt[3]{12}}{\sqrt[3]{2}} = \sqrt[3]{6}$

EXAMPLE 9. $\sqrt{10/25} = \dfrac{\sqrt{10}}{\sqrt{25}} = \dfrac{\sqrt{10}}{5}$ or $\frac{1}{5}\sqrt{10}$

We note that Property 2 may also be read in either direction.

Simplification of Radicals

We shall consider a radical to be in simplest form when:*

1. No factor can be removed from the radicand.
2. There are no fractions under the radical sign.
3. The radical does not appear in the denominator of a fraction.

---

* We may also simplify a radical by reducing the order.   For example, $\sqrt[4]{4} = \sqrt{2}$. We shall not consider this type of simplification.

By means of Properties 1 and 2, we are frequently able to write a radical in more simple form.

EXAMPLE 10. $\sqrt{20} = \sqrt{4} \cdot \sqrt{5} = 2\sqrt{5}$

EXAMPLE 11. $\sqrt[3]{16} = \sqrt[3]{8} \cdot \sqrt[3]{2} = 2\sqrt[3]{2}$

EXAMPLE 12. $\sqrt{2/5} = \sqrt{\dfrac{2}{5} \cdot \dfrac{5}{5}} = \dfrac{\sqrt{10}}{\sqrt{25}} = \dfrac{\sqrt{10}}{5}$ or $\frac{1}{5}\sqrt{10}$

EXAMPLE 13. $\sqrt[3]{1/2} = \sqrt[3]{\dfrac{1}{2} \cdot \dfrac{4}{4}} = \dfrac{\sqrt[3]{4}}{\sqrt[3]{8}} = \dfrac{\sqrt[3]{4}}{2}$ or $\frac{1}{2}\sqrt[3]{4}$

EXAMPLE 14. $\dfrac{\sqrt{3}}{\sqrt{7}} = \dfrac{\sqrt{3}}{\sqrt{7}}\dfrac{\sqrt{7}}{\sqrt{7}} = \dfrac{\sqrt{21}}{7}$ or $\frac{1}{7}\sqrt{21}$

The process illustrated in Examples 12, 13, and 14, is called *rationalizing the denominator*.   To see that for some purposes we actually simplify a radical by this process, compare the computation involved in evaluating the equivalent radicals $1/\sqrt{2}$ and $\sqrt{2}/2$ for the approximate value of $\sqrt{2}$ equal to 1.414.

Operations with Radicals

The two properties of radicals previously stated enable us to perform the fundamental operations with radicals.   We shall consider only radicals in which the radicand is a specific number.

Addition of Radicals

Two or more radicals can be added only if they have the same radicand and the same index; that is, if they are similar radicals.   In symbols

$$a\sqrt[n]{c} + b\sqrt[n]{c} = (a + b)\sqrt[n]{c}$$

This rule also applies to the subtraction of radicals.

EXAMPLE 15. $12\sqrt{3} - 4\sqrt{3} + 2\sqrt{5} - 8\sqrt{5} = 8\sqrt{3} - 6\sqrt{5}$

When the radicals are not similar, we frequently can make them similar by proper simplification.

EXAMPLE 16. $2\sqrt{2} + 4\sqrt{12} + 5\sqrt{8} - 2\sqrt{3} = 2\sqrt{2} + 8\sqrt{3}$
$+ 10\sqrt{2} - 2\sqrt{3} = 12\sqrt{2} + 6\sqrt{3}$

Multiplication of Radicals

Radicals having the same index may be multiplied by the use of Property 1.

EXAMPLE 17. $(4 \sqrt{2})(3 \sqrt{12}) = 12 \sqrt{24} = 24 \sqrt{6}$

EXAMPLE 18. $(2 + 3 \sqrt{2})(2 - 3 \sqrt{2}) = 4 - 6 \sqrt{2} + 6 \sqrt{2}$
$$- 9 \sqrt{4} = 4 - 18 = -14$$

Division of Radicals

Radicals having the same index may be divided by the use of Property 2. If the radicand in the denominator is not an exact divisor of the radicand in the numerator, we use the process of rationalizing the denominator to obtain the quotient in simplest form.

EXAMPLE 19. $\dfrac{\sqrt{8}}{\sqrt{2}} = \sqrt{4} = 2$

EXAMPLE 20. $\dfrac{\sqrt{5}}{\sqrt{3}} = \sqrt{\dfrac{5}{3}} = \sqrt{\dfrac{5}{3} \cdot \dfrac{3}{3}} = \dfrac{\sqrt{15}}{3}$ or $\tfrac{1}{3} \sqrt{15}$

or $\dfrac{\sqrt{5}}{\sqrt{3}} = \dfrac{\sqrt{5} \sqrt{3}}{\sqrt{3} \sqrt{3}} = \dfrac{\sqrt{15}}{3}$, as before

EXAMPLE 21. $\dfrac{4}{\sqrt{3}} = \dfrac{4 \sqrt{3}}{\sqrt{3} \sqrt{3}} = \dfrac{4 \sqrt{3}}{3}$ or $\tfrac{4}{3} \sqrt{3}$

EXAMPLE 22. $\dfrac{2 + \sqrt{3}}{2 - \sqrt{3}}.$ We may rationalize the denominator by multiplying numerator and denominator by the radical expression $2 + \sqrt{3}$. Therefore

$$\frac{2 + \sqrt{3}}{2 - \sqrt{3}} = \frac{(2 + \sqrt{3})(2 + \sqrt{3})}{(2 - \sqrt{3})(2 + \sqrt{3})} = \frac{7 + 4 \sqrt{3}}{4 - 3} = 7 + 4 \sqrt{3}$$

Approximate Value of an Expression Containing Radicals

Table I gives the approximate square roots and cube roots of numbers from 1 to 100. Of course tables are available for numbers greater than 100. However, by means of Table I we may approximate the numerical value of many expressions containing radicals. The radicals are usually written in simplest form before using the tables.

EXAMPLE 23. $2 + \sqrt{24} = 2 + 4.899 = 6.899$. (Since $\sqrt{24}$ can be obtained directly from Table I, it is not necessary to simplify this radical).

EXAMPLE 24. $\dfrac{4}{\sqrt{3}} = \dfrac{4\sqrt{3}}{3} = \dfrac{4(1.732)}{3} = 2.309$

EXAMPLE 25. $\dfrac{2 + \sqrt{8}}{2} = \dfrac{2 + 2\sqrt{2}}{2} = 1 + \sqrt{2} = 1 + 1.414 = 2.414$

Exercise 25

OPERATIONS WITH RADICALS

Which of the following numbers are rational?

1. 1.25 2. 0.65 3. 0.015 4. $1\frac{3}{5}$ 5. $0.666\ldots$
6. $\sqrt{2/3}$ 7. $\sqrt{25/4}$ 8. $\sqrt{3/5}$ 9. $1\frac{9}{16}$ 10. $\sqrt{25/9}$
11. $\sqrt{3/25}$ 12. $\sqrt{9\frac{2}{5}}$ 13. $\sqrt{1\frac{1}{2}}$ 14. $\sqrt{1\frac{11}{25}}$ 15. $\sqrt{0.01}$
16. $\sqrt{0.10}$

Simplify the following radicals by using Property 1:

17. $\sqrt{24}$ 18. $\sqrt{98}$ 19. $\sqrt{72}$ 20. $\sqrt{500}$ 21. $\sqrt{50}$
22. $5\sqrt{20}$ 23. $3\sqrt{28}$ 24. $\sqrt{162}$ 25. $4\sqrt{12}$ 26. $2\sqrt{147}$
27. $2\sqrt{80}$ 28. $\frac{1}{5}\sqrt{75}$ 29. $\sqrt[3]{16}$ 30. $2\sqrt[3]{24}$ 31. $2\sqrt[3]{54}$
32. $2\sqrt[3]{81}$ 33. $5\sqrt[3]{40}$ 34. $\frac{1}{2}\sqrt[3]{56}$ 35. $\frac{2}{3}\sqrt[3]{108}$ 36. $\frac{1}{3}\sqrt[3]{54}$

Simplify the following radicals by using Property 2:

37. $\sqrt{1/2}$ 38. $\sqrt{2/3}$ 39. $\sqrt{1/5}$ 40. $\sqrt{1/7}$ 41. $\sqrt{3/8}$
42. $\sqrt{2/5}$ 43. $\sqrt{5/6}$ 44. $\sqrt{2/11}$ 45. $\sqrt[3]{1/4}$ 46. $\sqrt[3]{1/9}$
47. $\sqrt[3]{1/16}$ 48. $\sqrt[3]{1/2}$ 49. $\sqrt[3]{3/4}$ 50. $\sqrt[3]{2/25}$ 51. $\sqrt[3]{2/5}$
52. $\sqrt[3]{4/5}$

Simplify, when necessary, the following radicals and combine similar terms:

53. $\sqrt{2} + 4\sqrt{12} - 3\sqrt{4}$ 54. $\sqrt{12} + \sqrt{27} + \sqrt{48}$
55. $2\sqrt{32} + 5\sqrt{18}$ 56. $4\sqrt{40} + \sqrt{250}$
57. $3\sqrt{75} + \sqrt{27} + 2\sqrt{12}$ 58. $2\sqrt{12} + 3\sqrt{8} - 2\sqrt{27}$
59. $2\sqrt{20} - 2\sqrt{32} + \sqrt{45}$ 60. $3\sqrt{12} - \sqrt{24} + \sqrt{54} - \sqrt{75}$
61. $2\sqrt[3]{16} - \sqrt[3]{54}$ 62. $4\sqrt[3]{108} - 2\sqrt[3]{192}$

Multiply the following radicals and simplify when possible:

63. $\sqrt{2} \cdot \sqrt{3}$ 64. $\sqrt{2}\sqrt{2}$ 65. $\sqrt{3}\sqrt{6}$

66. $2\sqrt{3} \cdot 3\sqrt{2}$ 67. $\sqrt{7}\sqrt{7}$ 68. $\sqrt{2}\sqrt{4}\sqrt{8}$

69. $4\sqrt{13} \cdot 2\sqrt{2}$ 70. $3\sqrt{3} \cdot 2\sqrt{30}$ 71. $2\sqrt[3]{16} \cdot 3\sqrt[3]{4}$

72. $5\sqrt[3]{81} \cdot 2\sqrt[3]{2}$ 73. $\sqrt[3]{7}\sqrt[3]{7}\sqrt[3]{7}$ 74. $\sqrt[3]{4}\sqrt[3]{50}\sqrt[3]{5}$

75. $\sqrt{3}\,(\sqrt{2} + \sqrt{3} + \sqrt{4})$

76. $2\sqrt{3}\,(\sqrt{3} - \sqrt{2})$

77. $(\sqrt{2} - \sqrt{10} - \sqrt{20})\,\sqrt{5}$

Find each of the following products, using the methods of Section 20:

78. $(\sqrt{2} + 5)(\sqrt{2} - 5)$ 79. $(\sqrt{3} - 2)(\sqrt{3} + 2)$

80. $(2\sqrt{2} - 3)(2\sqrt{2} + 3)$ 81. $(2 + 2\sqrt{3})(2 - 2\sqrt{3})$

82. $(\sqrt{2} + \sqrt{3})(\sqrt{2} - \sqrt{3})$ 83. $(\sqrt{2} - \sqrt{5})(\sqrt{2} - \sqrt{5})$

84. $(\sqrt{2} + 2\sqrt{3})(\sqrt{2} - 3\sqrt{3})$

85. $(2\sqrt{5} + 2\sqrt{3})(2\sqrt{5} - 2\sqrt{3})$

86. $(\sqrt{3} - 2\sqrt{7})(\sqrt{3} + 2\sqrt{7})$

87. $(\sqrt{3} - 2\sqrt{7})^2$

Find each of the following quotients, writing the results in simplest form:

88. $\dfrac{\sqrt{28}}{\sqrt{7}}$ 89. $\dfrac{\sqrt{42}}{\sqrt{21}}$ 90. $\dfrac{\sqrt{6}}{\sqrt{2}}$ 91. $\dfrac{\sqrt{3}}{\sqrt{2}}$ 92. $\dfrac{\sqrt{2}}{\sqrt{3}}$

93. $\dfrac{3\sqrt{5}}{2\sqrt{3}}$ 94. $\dfrac{3\sqrt{7}}{2\sqrt{3}}$ 95. $\dfrac{5\sqrt{15}}{4\sqrt{12}}$ 96. $\dfrac{2}{\sqrt{3}}$ 97. $\dfrac{2}{\sqrt{2}}$

98. $\dfrac{3}{\sqrt{3}}$ 99. $\dfrac{1}{\sqrt{2}}$ 100. $\dfrac{\sqrt[3]{16}}{\sqrt[3]{2}}$ 101. $\dfrac{\sqrt[3]{54}}{\sqrt[3]{2}}$ 102. $\dfrac{\sqrt[3]{12}}{\sqrt[3]{4}}$

103. $\dfrac{\sqrt[3]{16}}{\sqrt[3]{4}}$

Simplify each of the following expressions:

104. $\dfrac{4 + \sqrt{8}}{2}$ 105. $\dfrac{4 - \sqrt{12}}{2}$ 106. $\dfrac{6 - \sqrt{27}}{3}$

107. $\dfrac{6 - \sqrt{54}}{3}$ 108. $\dfrac{6 + \sqrt{54}}{3}$ 109. $\dfrac{6 + \sqrt{8}}{6}$

110. $\dfrac{8 + \sqrt{32}}{16}$ 111. $\dfrac{5 + \sqrt{200}}{10}$ 112. $\dfrac{4 - \sqrt{8}}{6}$

Using the table of square roots, evaluate the following expressions to three decimal places. Simplify the expressions first.

113. $\dfrac{2}{\sqrt{2}}$ **114.** $\dfrac{1}{\sqrt{3}}$ **115.** $\dfrac{2}{\sqrt{3}}$ **116.** $\dfrac{7}{\sqrt{7}}$ **117.** $\dfrac{\sqrt{2}}{\sqrt{3}}$

118. $\dfrac{\sqrt{10}}{\sqrt{3}}$ **119.** $\dfrac{\sqrt{6}}{\sqrt{2}}$ **120.** $\dfrac{\sqrt{3}}{\sqrt{2}}$ **121.** $\dfrac{4+\sqrt{8}}{2}$

122. $\dfrac{4-\sqrt{6}}{2}$ **123.** $\dfrac{3+\sqrt{2}}{2}$ **124.** $\dfrac{3-\sqrt{2}}{2}$ **125.** $\dfrac{6-\sqrt{54}}{2}$

126. $\dfrac{1-\sqrt{6}}{2}$ **127.** $\dfrac{3-\sqrt{5}}{2}$ **128.** $\dfrac{3+\sqrt{2}}{2}$ **129.** $\dfrac{4-\sqrt{40}}{4}$

26 The Set of Real Numbers

The set of rational numbers and the set of irrational numbers comprise the set of *real numbers*. The relation of the real numbers to other sets of numbers we have considered is shown in the following diagram, in which each set is a subset of the one above it.

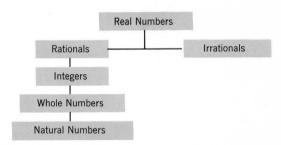

Figure 1

Decimal Representation of Real Numbers

By definition, each rational number can be expressed as a quotient of two integers, whereas each irrational number cannot be so expressed. This property leads to some interesting consequences when one attempts to express these numbers decimally. In what follows it is assumed that the reader is familiar with the ordinary uses of decimal notation.

Since a rational number is expressible as a quotient of two integers, its decimal representation can be obtained by merely performing the indicated division. One of two situations will result:

1. The decimal representation will *terminate*, as in $\frac{5}{16} = 0.3125$; or
2. The decimal representation will *repeat*, as in $\frac{5}{12} = 0.41666 \ldots$.

We can demonstrate that the converse is also true, that is, if a decimal terminates or develops a never-ending, repeating cycle, it can be expressed as the quotient of two integers. The procedure is illustrated in the following examples.

EXAMPLE 1.   Express the terminating decimal 0.435 as a quotient of two integers.

Solution. $0.435 = \dfrac{435}{1000} = \dfrac{107}{200}$

EXAMPLE 2.   What rational number is equivalent to the repeating decimal 0.6363 . . . ?

Solution.   Let $n = 0.6363 \ldots$
             then $100n = 63.6363 \ldots$
             Subtracting $n = 0.6363 \ldots$
             we have $99n = 63$
             therefore, $n = \dfrac{63}{99} = \dfrac{7}{11}$

EXAMPLE 3.   Show that the repeating decimal 3.222 . . . is rational.

Solution.   Let $n = 3.222 \ldots$
             then $10n = 32.222 \ldots$
             subtracting $n = 3.222 \ldots$
             we have $9n = 29$
             therefore $n = \dfrac{29}{9}$, which is rational

These examples suggest that a decimal numeral represents a rational number if and only if it terminates or repeats. It follows that the decimal representation of an irrational number can only be approximated since it will neither terminate nor develop a repeating cycle of digits. Thus the ten place decimal approximation of π, an irrational number, is 3.1415926536, and the five place decimal approximation of $\sqrt{2}$ is 1.41421. In neither case can an exact decimal representation be obtained, no matter how many places the computation is carried forward.

Graphic Representation of Real Numbers

We have noted in Section 4 that a one-to-one correspondence can be established between the set of integers and certain points on a line. If we divide each line segment determined by the integers into three equal parts, the points of division would correspond to the rational numbers $\ldots -\frac{4}{3}, -\frac{3}{3}, -\frac{2}{3}, -\frac{1}{3}, 0, \frac{1}{3}, \frac{2}{3}, \frac{3}{3}, \frac{4}{3}, \frac{5}{3}, \ldots$, as shown in Figure 2. We

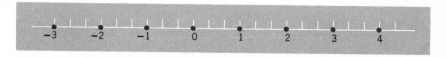

Figure 2

could, of course, divide each of the original segments into four equal parts, or eight, or, in general, n parts to set up a correspondence between points on the number scale and fractions of the form $\ldots -3/n, -2/n, -1/n, 0, 1/n, 2/n, 3/n, \ldots$. Clearly we could subdivide the given segments into smaller and smaller parts, with the result that the points of division would be closer and closer to each other.

It is difficult to imagine that, in this process, any point on the number scale could ultimately escape being associated with some rational number. However, our experience with irrational numbers indicates that this is so. For example, we know that no point arrived at in this manner can represent $\sqrt{2}$, since $\sqrt{2}$ is not expressible as a quotient of two integers. That a point corresponding to $\sqrt{2}$ on the number scale does exist is easily illustrated.

If a right triangle whose legs are each 1 unit long is constructed on the number scale as illustrated in Figure 3, the length of its hypotenuse is

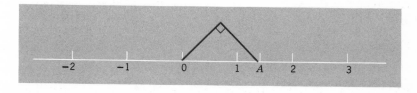

Figure 3

$\sqrt{2}$. Therefore the distance $0A$ corresponds to the number $\sqrt{2}$, and the point A is associated with that number. Similarly, each irrational number can be associated with a unique point on the number scale.

Thus each point on the number scale is placed in correspondence with either a rational or irrational number, and, conversely, each such number is placed in correspondence with a single point on the scale. This is equivalent to the statement that a one-to-one correspondence exists between the elements of the set of real numbers and the points on a line.

Properties of the Real Numbers

The extension of the number system from rationals to real numbers preserves all the properties developed in Section 9 for the set of rationals. Although the proof of this is beyond the scope of this book, it is important to recognize that, in particular, the eleven postulates for a number field apply to the set of real numbers.

Thus the set contains an additive identity, 0, a multiplicative identity, 1, an additive inverse for each element in the set, and a multiplicative inverse for each element in the set except 0. For example, the additive inverse of the element $\sqrt{3}$ is $-\sqrt{3}$, and the multiplicative inverse of $\sqrt{3}$ is $1/\sqrt{3}$ or $\sqrt{3}/3$. Further, the set is closed with respect to both addition and multiplication, that is, the sum and the product of two real numbers is always a real number.

Finally, the set of real numbers obeys the usual commutative, associative, and distributive postulates for addition and multiplication. The following examples are illustrations of these properties.

EXAMPLE 1. $\sqrt{3} \cdot \sqrt{2} = \sqrt{2} \cdot \sqrt{3}$ by the commutative postulate for multiplication.

EXAMPLE 2. $\frac{1}{2} + \sqrt{5} = \sqrt{5} + \frac{1}{2}$ by the commutative postulate for addition.

EXAMPLE 3. $(3 \cdot \sqrt{2}) \cdot \sqrt{8} = 3(\sqrt{2} \cdot \sqrt{8})$ by the associative postulate for multiplication.

EXAMPLE 4. $5 + (3 + \sqrt{13}) = (5 + 3) + \sqrt{13}$ by the associative postulate for addition.

EXAMPLE 5. $2(\sqrt{3} + \sqrt{6}) = 2\sqrt{3} + 2\sqrt{6}$ by the distributive postulate.

Although further extensions of the number system are possible, we will limit our work in algebra to the set of real numbers.   Therefore, unless otherwise specified, the word number as used hereafter in this book will mean real number.

Exercise 26

THE SET OF REAL NUMBERS

Given a line segment of length $\sqrt{2}$ on a number scale, mark the point corresponding to each of the following numbers and identify each as either rational or irrational:

1. $-2\sqrt{2}$ **2.** $\sqrt{2}+1$ **3.** $1-\sqrt{2}$ **4.** $2\sqrt{2}+1$

To how many of the following sets of numbers do each of the given numbers belong? U = real numbers, R = rationals, R' = irrationals, I = integers, W = whole numbers, N = natural numbers.

5. -2 **6.** $\frac{1}{2}$ **7.** 0 **8.** $\sqrt{5}$
9. 0.63 **10.** $\sqrt{81}$ **11.** $4.7171\ldots$ **12.** 193

Show that the following numbers can be expressed either as terminating or repeating decimals:

13. $\frac{7}{16}$ **14.** $\frac{1}{7}$ **15.** $\frac{31}{4}$ **16.** $\frac{27}{11}$

Show that each of the following numbers can be expressed as a quotient of two integers and is therefore rational:

17. 1.78 **18.** 0.11 **19.** $0.1313\ldots$
20. $5.555\ldots$ **21.** 0.003 **22.** $0.327327\ldots$

Since $\sqrt{5}$ is irrational, determine whether each of the following expressions is rational or irrational.

23. $2\sqrt{5}$ **24.** $3+\sqrt{5}$ **25.** $1-\sqrt{5}$
26. $(3+\sqrt{5})(3-\sqrt{5})$ **27.** $(3+\sqrt{5})+(3-\sqrt{5})$ **28.** $\sqrt{5}\cdot\sqrt{5}$

State a postulate which justifies each of the following equalities.

29. $3+\sqrt{2}=\sqrt{2}+3$ **30.** $3(2+\sqrt{3})=3\cdot2+3\sqrt{3}$
31. $\sqrt{2}\cdot\sqrt{5}=\sqrt{5}\cdot\sqrt{2}$ **32.** $\sqrt{7}+0=0+\sqrt{7}=\sqrt{7}$
33. $(2\cdot3)\sqrt{5}=2\cdot(3\sqrt{5})$
34. $(1+\sqrt{2})+\sqrt{3}=1+(\sqrt{2}+\sqrt{3})$

27   The Quadratic Formula

Although the left member of the quadratic equation $ax^2 + bx + c = 0$ may always be factored, in many cases the factors are rather complicated. This was the case in Section 24, where the factored form of $x^2 + 2x - 5 = 0$ was $(x + 1 - \sqrt{6})(x + 1 + \sqrt{6}) = 0$. Since factors of this type are not readily found, we seek a more convenient method of solving such equations.

The Quadratic Formula

We will now develop a formula for solving any quadratic equation. Our procedure in developing this formula is to rewrite the equation so that the left member is a perfect square. By taking the square root of both members of this equation, we reduce the quadratic equation to two linear equations which may then be solved in the usual way.

Let the equation be

$$ax^2 + bx + c = 0$$

Divide both members by a.

$$x^2 + \frac{b}{a}x + \frac{c}{a} = 0$$

Add $-\frac{c}{a}$ to both members.

$$x^2 + \frac{b}{a}x = -\frac{c}{a}$$

Add $\frac{b^2}{4a^2}$ to both members.

$$x^2 + \frac{b}{a}x + \frac{b^2}{4a^2} = \frac{b^2}{4a^2} - \frac{c}{a}$$

Write the left member as the square of a binomial and combine the terms of the right member.

$$\left(x + \frac{b}{2a}\right)^2 = \frac{b^2 - 4ac}{4a^2}$$

Take the square root of both members.

$$x + \frac{b}{2a} = + \frac{\sqrt{b^2 - 4ac}}{2a}$$

$$\text{or} \quad x + \frac{b}{2a} = - \frac{\sqrt{b^2 - 4ac}}{2a}$$

Add $-\dfrac{b}{2a}$ to both members.

$$x = -\frac{b}{2a} + \frac{\sqrt{b^2 - 4ac}}{2a}$$

$$\text{and} \quad x = -\frac{b}{2a} - \frac{\sqrt{b^2 - 4ac}}{2a}$$

The last two equations may be combined to give the equation

$$x = \frac{-b \pm \sqrt{b^2 - 4ac}}{2a}$$

which is the *quadratic formula*.

Using the Quadratic Formula

To solve a quadratic equation by means of the quadratic formula:

1. Write the equation in standard form.
2. Write the specific values of a, b, and c.
3. Substitute these values in the formula.
4. Simplify the result.
5. Find the roots by first using the plus sign and then the minus sign or vice versa.
6. Check each root in the original equation.

EXAMPLE 1.   Solve the equation $2x^2 - 5x + 3 = 0$.

(1) $2x^2 - 5x + 3 = 0$

(2) $a = 2, b = -5, c = 3$

(3) $x = \dfrac{5 \pm \sqrt{(-5)^2 - 4(2)(3)}}{2(2)}$

(4) $x = \dfrac{5 \pm \sqrt{1}}{4}$

(5) $x = \dfrac{5 + 1}{4} = \dfrac{3}{2}$, and $x = \dfrac{5 - 1}{4} = 1$

(6)   Substituting $x = \frac{3}{2}$ in the given equation,
$$2(\tfrac{3}{2})^2 - 5(\tfrac{3}{2}) + 3 = 0$$
$$0 = 0$$
Substituting $x = 1$ in the given equation,
$$2(1)^2 - 5(1) + 3 = 0$$
$$0 = 0$$

EXAMPLE 2.   Solve the equation $x^2 = 4x - 1$.

(1) $x^2 - 4x + 1 = 0$

(2) $a = 1, b = -4, c = 1$

(3) $x = \dfrac{-(-4) \pm \sqrt{(-4)^2 - 4(1)(1)}}{2(1)}$

(4) $x = \dfrac{4 \pm \sqrt{12}}{2} = 2 \pm \sqrt{3}$

(5) $x = 2 + \sqrt{3}$ and $x = 2 - \sqrt{3}$

(6)   If these values of x are substituted in the original equation it will be found that the roots satisfy the given equation, so that $x = 2 + \sqrt{3}$ and $x = 2 - \sqrt{3}$ are roots of the equation.

If $\sqrt{3}$ is taken to be 1.732, the roots are approximately $(2 + 1.732)$ $= 3.732$ and $(2 - 1.732) = 0.268$.   The equation may be checked with these values of the roots, but the student should bear in mind that since these values are approximations, the check may not be exact.

The Discriminant: $b^2 - 4ac$

In the quadratic formula

$$x = \frac{-b \pm \sqrt{b^2 - 4ac}}{2a}$$

the quantity $b^2 - 4ac$ is called the *discriminant*, since it enables us to determine the character of the roots of the equation without solving it. In this discussion, a, b, and c are limited to the set of rational numbers. Let us first suppose that a, b, and c, in the general quadratic equation $ax^2 + bx + c = 0$, have such values that $b^2 - 4ac = 0$.   Then we shall have, from the quadratic formula,

$$x = \frac{-b \pm \sqrt{0}}{2a} = \frac{-b}{2a}$$

In this case, the roots of the equation are equal and rational.

EXAMPLE 3.   Determine the character of the roots of the equation

$$4x^2 - 12x + 9 = 0$$

The value of $b^2 - 4ac = (-12)^2 - 4(4)(9) = 144 - 144 = 0$.   Therefore the roots are equal and rational.   This may, of course, be verified by solving the equation, the roots being $\frac{3}{2}, \frac{3}{2}$.

Let us suppose that $b^2 - 4ac$ is a quantity greater than zero.   In this case, it is clear that the roots are real and unequal since we first add and then subtract a quantity greater than zero.

EXAMPLE 4.   Determine the character of the roots of the equation

$$2x^2 + 19x + 35 = 0$$

The value of $b^2 - 4ac = (19)^2 - (4)(2)(35) = 81$.   Therefore the roots are real and unequal, which may be verified as in Example 3 by actually solving the equation.

It should be noted that the value of the discriminant in Example 4 is 81, a perfect square.   We may make the additional observation that when $b^2 - 4ac$ is a perfect square, the roots are rational numbers; if $b^2 - 4ac$ is not a perfect square, the roots are irrational numbers.

Finally, let us suppose that $b^2 - 4ac$ is a quantity less than zero.   Then $\sqrt{b^2 - 4ac}$ is called an *imaginary number*.   Expressions such as $3 + \sqrt{-2}$ are called complex numbers, so that when $b^2 - 4ac$ is negative, the roots are complex numbers.

EXAMPLE 5.   Determine the character of the roots of the equation

$$x^2 + 3x + 5 = 0$$

The value of $b^2 - 4ac = 3^2 - (4)(1)(5) = -11$.   Therefore the roots are complex numbers, which may be verified as in Example 3 by actually solving the equation.

The above discussion may be summarized as follows:

In the quadratic equation $ax^2 + bx + c = 0$, where a, b, and c are rational numbers:

1. If $b^2 - 4ac = 0$, the roots of the equation are rational and equal.
2. If $b^2 - 4ac > 0$, the roots of the equation are real and unequal. They are rational if $b^2 - 4ac$ is a perfect square, and irrational otherwise.
3. If $b^2 - 4ac < 0$, the roots are complex.

Exercise 27

THE QUADRATIC FORMULA

Solve the following equations.   (Remember that checking an equation is a necessary part of its solution.)

1. $2x^2 - 5x + 3 = 0$ 2. $3x^2 + 5x - 2 = 0$
3. $2x^2 - 9x + 4 = 0$ 4. $6x^2 - 7x + 2 = 0$
5. $6x^2 + 19x + 10 = 0$ 6. $6x^2 = x + 1$
7. $4x^2 = 6 - 5x$ 8. $4x = 1 - 5x^2$
9. $2x^2 = 10 - x$ 10. $2x^2 - 7x = 15$
11. $4x^2 - 4x - 3 = 0$ 12. $4x^2 - 8x + 3 = 0$
13. $9x^2 - 9x + 2 = 0$ 14. $2x^2 + 5x + 2 = 0$
15. $5x^2 + 18x - 8 = 0$ 16. $6x^2 - 5x - 6 = 0$
17. $x^2 + x - 6 = 0$ 18. $6x^2 - 13x + 6 = 0$
19. $10x^2 - 9x + 2 = 0$

Solve the following equations. Evaluate the roots to three decimal places.

20. $x^2 + x - 1 = 0$ 21. $2x^2 + 2x - 1 = 0$ 22. $x^2 + x - 3 = 0$
23. $2x^2 + x - 4 = 0$ 24. $x^2 + 3x + 1 = 0$ 25. $2x^2 - 8x + 3 = 0$
26. $2x^2 + 5x + 1 = 0$ 27. $x^2 + 4x + 2 = 0$ 28. $3x^2 - 8x + 2 = 0$
29. $4x^2 - 6x + 1 = 0$ 30. $5x^2 + 5x - 2 = 0$

Determine, without solving, the character of the roots of each of the following equations.

31. $x^2 - 7x + 4 = 0$ 32. $6x^2 + x - 35 = 0$ 33. $2x^2 - 3x + 5 = 0$
34. $3x^2 + x - 2 = 0$ 35. $x^2 - x + 1 = 0$ 36. $x^2 - 6x + 9 = 0$
37. $4x^2 + 4x + 1 = 0$ 38. $9x^2 - 30x = -25$ 39. $4x^2 + 4x - 1 = 0$
40. $x^2 + x + 2 = 0$

28   Applications of Quadratic Equations

In Section 16, the symbolic form of the problems solved was always a linear equation. We now consider problems whose symbolic form is a quadratic equation. In general, the process of "setting up the equation" for problems involving quadratic equations is the same as for problems involving linear equations. Since every quadratic equation has two roots, it is very important that the apparent roots be checked for their meaning in relation to the stated problem, even if they do satisfy the equation.

EXAMPLE 1.   The length of a rectangle is 5 inches more than the width. Find the dimensions if the area is 24 square inches.

Let x = width in inches.   Then $x + 5$ = length in inches and $x(x + 5)$ = 24 (since area equals length times width).

$$x^2 + 5x - 24 = 0$$
$$(x + 8)(x - 3) = 0$$
$$x = 3, x = -8$$

Obviously, the root $x = -8$ must be discarded, since -8 inches as the width of a rectangle is meaningless.   Therefore the width is 3 inches, and the length is $x + 5$ or 8 inches.   The area is then 24 square inches as required.

EXAMPLE 2.   Find two numbers whose sum is 29 and whose product is 198.

Let n = one of the numbers; then, $29 - n$ = other number.   Since their product is 198, we have

$$n(29 - n) = 198$$
$$n^2 - 29n + 198 = 0$$
$$(n - 11)(n - 18) = 0$$
$$n = 11, n = 18$$
If $n = 11$, then $(29 - n) = 18$
If $n = 18$, then $(29 - n) = 11$

Therefore the two numbers are 18 and 11.   Note that both roots of the quadratic equation are admissible values of one of the required numbers.   However, due to a duplication of results, there is only one pair of values which satisfies the stated problem.

Exercise 28

APPLICATIONS OF QUADRATIC EQUATIONS

1. Four times the square of a number equals twenty-five times the number.   Find the number.

2. The square of a positive number plus the number itself equals 56. Find the number.

3. The square of a number plus three times the number equals 40. Find the number.

4. The difference between two numbers is 6, and the difference between their squares is 216.   Find the numbers.

5. The sum of the squares of two consecutive integers is 761. Find the integers.

6. The area of a rectangular living room is 264 square feet. The length is 10 feet longer than the width. Find the dimensions of the room.

7. The area of a rectangle is 540 square feet. The perimeter is 94 feet. Find the dimensions of the rectangle.

8. The total surface of a cube is 150 square inches. Find the edge of the cube.

9. The base of a triangle is 7 inches longer than the altitude. The area is 60 square inches. Find the base and altitude. *Hint:* $A = \frac{1}{2}bh$.

10. A plot of grass is three times as long as it is wide. The grass is enclosed by a sidewalk 10 feet wide. The area of the grass is 1200 square feet more than the area of the sidewalk. Find the area of the plot of grass.

11. The hypotenuse of a right triangle is 10 inches, and one of the legs is 2 inches longer than the other. Find the length of the legs. *Hint:* If a and b are the legs of the triangle, and c is the hypotenuse, then $a^2 + b^2 = c^2$.

12. The three sides of a right triangle are consecutive integers. Find the sides of the triangle.

13. A tray is formed by cutting a square from each corner of a rectangular sheet of tin 5 inches by 7 inches and bending up the edges. If the base area is to be 29 square inches, what is the approximate size of the corners cut off?

14. A square sheet of tin has $\frac{1}{2}$-inch squares cut from each corner. When this piece is folded to form a tray, the volume is 40.5 cubic inches. What are the dimensions of the original piece of tin?

The height h to which a ball thrown vertically upward will rise in t seconds, when thrown with a velocity of 100 feet per second, is given approximate y by the formula $h = 100t - 16t^2$.

15. In how many seconds will the ball rise to 144 feet?

16. What is the significance of the other root of the equation?

17. In how many seconds will the ball reach the ground?

18. Assuming that the time to reach the greatest height is one-half the total time in the air, in how many seconds will the ball reach the greatest height?

19. What is the greatest height the ball will reach?

20. If a ball is thrown vertically upward with a velocity of 80 feet per second, from a height of h feet above the ground, the time to return to the ground is given approximately by the formula $16t^2 - 80t - h = 0$. Find the time to reach the ground if $h = 96$ feet.

21. The radius of a right circular cylinder is 12 inches and the altitude is 2 inches. What additional length added to either altitude or radius will produce the same increase in volume? *Hint:* The volume of a right circular cylinder is $\pi r^2 h$.

22. Interpret both roots of the equation of Problem 21.

23. The greatest distance in miles a pilot can see when at a height of h miles above the ground is approximately given by the formula $d^2 = h^2 + 800h$. Find the height to which the plane must rise in order for the pilot to see a distance of 10 miles.

24. An approximate formula for the relation between the distance d feet in which a car going r miles an hour can be stopped is $d = 0.045r^2 + 1.1r$. What is the maximum speed a car can travel in order for the driver to avoid hitting a child 30 feet away?

25. When the length l and width w of a rectangle are related by the equation $l^2 - lw - w^2 = 0$, the proportions are considered to be most pleasing to the eye. What is the length of a rectangle of these proportions if the width is 5 feet?

26. The formula for the side s of a regular decagon inscribed in a circle of radius r is $s^2 + rs - r^2 = 0$. Find the side of a decagon inscribed in a circle of 10 inch radius.

CHAPTER **6**

Fractions and the Solution
of Fractional Equations

Algebraic Fractions

If a and b are any two algebraic expressions, and b is not equal to zero, then a/b is an algebraic fraction. As in the numerical fractions of arithmetic, a is the numerator and b is the denominator. We also call a and b the terms of the fraction. Examples of algebraic fractions are:

$$\frac{x+2}{x-3}, \quad \frac{2x}{x+2}, \quad \frac{x}{y}, \quad \frac{x^2 - 2xy + y^2}{2x - y}$$

Since a fraction whose denominator equals zero is meaningless, we say that the fraction $(x + 2)/(x - 3)$ is *undefined* for $x = 3$, since $3 - 3 = 0$, and *defined* for all other values of x. Similarly, the fraction $2x/(x + 2)$ is undefined for $x = -2$, and defined for all other values of x.

The Fundamental Principle of Fractions

As in arithmetic, the basis of all operations with fractions is the following principle:

The value of a fraction is unchanged when the numerator and denominator are multiplied or divided by the same expression, provided the expression does not equal zero.

136

Equivalent Fractions

Fractions which have been obtained by use of the preceding principle are called *equivalent fractions*, since they have the same value. The arithmetic fractions $\frac{1}{2}$, $\frac{2}{4}$, $\frac{4}{8}$, $\frac{8}{16}$ are equivalent fractions, since the last three were obtained by multiplying both terms of $\frac{1}{2}$ by 2, 4, and 8, respectively, and each fraction equals $\frac{1}{2}$. The following pairs of algebraic fractions are equivalent; the multiplier or divisor for each pair is indicated in parentheses.

(a) $\dfrac{x}{y}$, $\dfrac{2x}{2y}$ $\qquad\qquad$ (2)

(b) $\dfrac{x+y}{x-y}$, $\dfrac{xy+y^2}{xy-y^2}$ $\qquad\qquad$ (y)

(c) $\dfrac{2ax+b}{3ax+b}$, $\dfrac{4ax+2b}{6ax+2b}$ $\qquad\qquad$ (2)

(d) $\dfrac{x^2-y^2}{x^2-2xy+y^2}$, $\dfrac{x+y}{x-y}$ $\qquad\qquad$ $(x-y)$

Reduction of Fractions to Lowest Terms

When the numerator and denominator of a fraction have no common factor, the fraction is said to be expressed in its lowest terms. A fraction which is not expressed in its lowest terms may be reduced to its lowest terms by dividing both numerator and denominator by the common factor. To avoid errors in reducing fractions, it is advisable to factor both numerator and denominator before dividing by the common factor.

EXAMPLE 1. $\dfrac{2x+4y}{4x+12y} = \dfrac{\cancel{2}(x+2y)}{\underset{2}{\cancel{4}}(x+3y)} = \dfrac{(x+2y)}{2(x+3y)}$

EXAMPLE 2. $\dfrac{x^2+7x-18}{x^2-3x+2} = \dfrac{(x+9)\cancel{(x-2)}}{(x-1)\cancel{(x-2)}} = \dfrac{x+9}{x-1}$

EXAMPLE 3. $\dfrac{x}{x^2+xy} = \dfrac{\cancel{x}}{\cancel{x}(x+y)} = \dfrac{1}{x+y}$

EXAMPLE 4. $\dfrac{64m^2}{128m^4} = \dfrac{\cancel{64m^2}}{2m^2\cancel{(64m^2)}} = \dfrac{1}{2m^2}$

Sometimes a student further attempts to reduce the result in a problem such as Example 2 by writing

$$\frac{x+9}{x-1} = \frac{\cancel{x}+9}{\cancel{x}-1} = \frac{9}{-1} = -9$$

The result is incorrect because x is not a factor of either the numerator or the denominator. The student has not divided both terms of the fraction by x; he has subtracted x from both terms of the fraction, which is not a permissible operation. If one writes

$$\frac{11}{12} = \frac{\cancel{10}+1}{\cancel{10}+2} = \frac{1}{2}$$

the error is at once apparent.

The Signs of a Fraction

Every fraction has three signs associated with it; the sign of the numerator, the sign of the denominator, and the sign before the fraction. If the sign is unwritten, it is considered to be positive. Consider the fraction a/b; by the fundamental principle of fractions its value is unchanged if we multiply both numerator and denominator by -1, so that we may write

$$\frac{a}{b} = \frac{-a}{-b}$$

Applying the same principle to the fraction $-a/+b$, we have

$$\frac{-a}{+b} = \frac{+a}{-b}$$

It also follows, from the rule for division of directed numbers, that

$$\frac{-a}{b} = -\frac{a}{b}$$

and that

$$\frac{a}{-b} = -\frac{a}{b}$$

These considerations lead us to the following principle:

The value of a fraction is unchanged if any two of its three signs are changed.

Thus, applying this principle, we have

$$\frac{6}{2} = \frac{-6}{-2} = -\frac{6}{-2} = -\frac{-6}{2}$$

which may be verified by noting that each fraction equals $+3$.

To apply this principle to algebraic fractions, it should be noted that (1) the sign of the numerator or denominator of a fraction is changed by changing the sign of every term of the numerator or denominator, and (2) the sign of the numerator or denominator of a fraction is changed by changing the sign of any factor of the numerator or denominator.

These principles are illustrated in the following examples:

EXAMPLE 5. $\dfrac{a-b}{b-a} = -\dfrac{(a-b)}{(a-b)} = -1$

EXAMPLE 6. $\dfrac{x+y-2}{-x-y+2} = -\dfrac{x+y-2}{x+y-2} = -1$

EXAMPLE 7. $-\dfrac{(a+b)(2a-b)}{(2a+b)(b-2a)} = \dfrac{(a+b)(2a-b)}{(2a+b)(2a-b)} = \dfrac{a+b}{2a+b}$

EXAMPLE 8. $\dfrac{(x+y)(y-x)}{(x-y)^2} = -\dfrac{(x+y)(x-y)}{(x-y)(x-y)} = -\dfrac{x+y}{x-y}$

or $+\dfrac{x+y}{y-x}$

Exercise 29

EQUIVALENT FRACTIONS

State the value or values of x (if any) for which each of the following fractions are undefined:

1. $\dfrac{x+2}{x-2}$ 2. $\dfrac{x+a}{x+b}$ 3. $\dfrac{x+a}{x-b}$

4. $\dfrac{x+1}{x^2-4}$ 5. $\dfrac{x+1}{x^2+4}$ 6. $\dfrac{x-y}{x+5}$

7. $\dfrac{x-2}{x^2-2x+1}$ 8. $\dfrac{3}{x^2-3x-10}$ 9. $\dfrac{x^2+1}{2x^2+5}$

Change each fraction to an equivalent fraction with indicated numerator or denominator:

10. $\dfrac{1}{x} = \dfrac{}{xy}$ **11.** $\dfrac{x}{2} = \dfrac{x^2}{}$ **12.** $\dfrac{2x}{3} = \dfrac{}{12x^2}$

13. $\dfrac{x}{y} = \dfrac{}{xyz}$ **14.** $\dfrac{x^2y}{3} = \dfrac{}{6}$ **15.** $\dfrac{c}{\pi} = \dfrac{}{\pi r^2}$

16. $\dfrac{1}{x+y} = \dfrac{}{x^2-y^2}$ **17.** $\dfrac{x}{x+y} = \dfrac{x^2-xy}{}$ **18.** $\dfrac{x+y}{x-y} = \dfrac{}{x^2-y^2}$

19. $\dfrac{x-y}{x} = \dfrac{}{xy}$ **20.** $\dfrac{x-y}{x} = \dfrac{x^2-2xy+y^2}{}$

21. $\dfrac{x^2-2xy+y^2}{x^2-y^2} = \dfrac{}{x+y}$ **22.** $\dfrac{a^2-3a+2}{a^2-a-2} = \dfrac{}{a+1}$

Reduce each of the following fractions to lowest terms:

23. $\dfrac{10x^2}{25x^3}$ **24.** $\dfrac{64m^2}{16m}$ **25.** $\dfrac{3ax}{21a^2}$

26. $\dfrac{12x}{24x^2}$ **27.** $\dfrac{12x^2}{2x}$ **28.** $\dfrac{28ax}{32a^2x^2}$

29. $\dfrac{a^3b}{a^2b^4}$ **30.** $\dfrac{15ab^3}{25a^2b^2x}$ **31.** $\dfrac{42x^4y^4z}{56x^3y^5z}$

32. $\dfrac{3x}{3x+3}$ **33.** $\dfrac{15n}{5n^2-15n}$ **34.** $\dfrac{2x}{4x^2-8x^3}$

35. $\dfrac{x^2-2x+1}{x^2-1}$ **36.** $\dfrac{n^3-4n}{n^2+4n+4}$ **37.** $\dfrac{3ax+3a^2}{3a^4}$

38. $\dfrac{x^2-x-6}{2x^2+x-6}$ **39.** $\dfrac{x^2-16}{x^2+x-20}$ **40.** $\dfrac{x^2-5x+6}{x^2-4x+4}$

41. $\dfrac{ax-ay}{x^2-y^2}$ **42.** $\dfrac{3x-9}{2x-6}$ **43.** $\dfrac{2y^2-y-10}{y^2-2y-8}$

44. $\dfrac{x^2-5xy+4y^2}{x^2-16y^2}$ **45.** $\dfrac{x^2-5x+6}{2x^2-12x+18}$

Write the proper sign, $+$ or $-$, in front of each equivalent fraction:

46. $\dfrac{x+y}{x-y} = (\quad)\dfrac{x+y}{y-x}$ **47.** $\dfrac{a-2b}{a-b} = (\quad)\dfrac{2b-a}{b-a}$

48. $-\dfrac{x+2}{2-x} = (\quad)\dfrac{x+2}{x-2}$ **49.** $\dfrac{x-y+2}{y-x} = (\quad)\dfrac{-x+y-2}{x-y}$

50. $\dfrac{x+y}{y+2x} = (\quad)\dfrac{x+y}{2x+y}$ **51.** $\dfrac{2x-y+3}{x-y-3} = (\quad)\dfrac{2x-y+3}{-3+x-y}$

52. $\dfrac{(x-y)(x-2)}{(x-2y)(x-3)} = (\quad)\dfrac{(y-x)(2-x)}{(2y-x)(3-x)}$

Reduce each of the following fractions, where possible, to lowest terms:

53. $\dfrac{a-b}{2a-2b}$ **54.** $-\dfrac{x-y}{y-x}$ **55.** $\dfrac{x-y}{y+x}$

56. $\dfrac{(x-2)(x-3)}{(2-x)(x+4)}$ **57.** $\dfrac{x+y-2}{2-x-y}$ **58.** $\dfrac{x+y-2}{x+y+2}$

59. $\dfrac{a^2-ab}{2b-2a}$ **60.** $\dfrac{x^2y-xy^2}{y^2-xy}$ **61.** $\dfrac{n^2-4}{2-n}$

62. $\dfrac{a^2-2ab+b^2}{2(b-a)}$ **63.** $\dfrac{x^2-9}{3y-xy}$ **64.** $\dfrac{2a^2-2a-12}{9-a^2}$

65. $\dfrac{x+2}{x+1}$ **66.** $\dfrac{acx}{ac+cx}$ **67.** $\dfrac{2a}{a+2}$

68. $\dfrac{2a}{2a+2}$ **69.** $\dfrac{x}{x+1}$ **70.** $\dfrac{x}{x^2-x}$

71. If the fraction $\dfrac{(x-2)^2}{x+1}$ is changed to $\dfrac{(2-x)^2}{x+1}$, is the sign of the numerator changed? Explain your answer.

72. If the fraction $\dfrac{(x-2)^3}{x+1}$ is changed to $\dfrac{(2-x)^3}{x+1}$, is the sign of the numerator changed? Explain your answer.

73. Is the fraction $\dfrac{(x-2)^2}{(x+1)}$ equivalent to $\dfrac{(2-x)^2}{x+1}$ or $-\dfrac{(2-x)^2}{x+1}$?

74. Is the fraction $\dfrac{(x-2)^3}{x+1}$ equivalent to $\dfrac{(2-x)^3}{x+1}$ or $-\dfrac{(2-x)^3}{x+1}$?

30 Multiplication and Division of Fractions

The Product of Two Fractions

As defined in Section 9 for rational numbers, if a/b and c/d are any two algebraic fractions, then

$$\frac{a}{b}\cdot\frac{c}{d}=\frac{ac}{bd}$$

Expressed in words, the product of two (or more) fractions is equal to the product of their numerators divided by the product of their denominators.

EXAMPLE 1. $\dfrac{3x}{2y}\cdot\dfrac{5x}{7y}=\dfrac{15x^2}{14y^2}$

EXAMPLE 2.

$$\frac{3a - 2}{3} \cdot \frac{2a - 1}{5} = \frac{(3a - 2)(2a - 1)}{15} = \frac{6a^2 - 7a + 2}{15}$$

Any factor which occurs in both numerator and denominator of the indicated product may be eliminated by division before the actual multiplication is done, so that the result will be in lowest terms.

EXAMPLE 3.

$$\frac{a^2 - x^2}{a} \cdot \frac{a}{a + x} = \frac{(a - x)(a + x)}{a} \cdot \frac{a}{a + x} = a - x$$

EXAMPLE 4.

$$\frac{x^2 - 3x - 4}{2 - x} \cdot \frac{x^2 - 4}{2x + 2} = -\frac{(x - 4)(x + 1)}{x - 2} \frac{(x - 2)(x + 2)}{2(x + 1)}$$
$$= -\frac{(x - 4)(x + 2)}{2} = -\frac{x^2 - 2x - 8}{2}$$

The Reciprocal of a Number

If a and b are any two numbers, or algebraic expressions, such that $ab = 1$, then a is the reciprocal of b and b is the reciprocal of a.

EXAMPLE 5.   Since $2 \cdot \frac{1}{2} = 1$, 2 is the reciprocal of $\frac{1}{2}$ and $\frac{1}{2}$ is the reciprocal of 2.

EXAMPLE 6.   Since $\frac{2}{3} \cdot \frac{3}{2} = 1$, $\frac{2}{3}$ is the reciprocal of $\frac{3}{2}$, and $\frac{3}{2}$ is the reciprocal of $\frac{2}{3}$.

EXAMPLE 7.   Since $\dfrac{a + b}{a - b} \cdot \dfrac{a - b}{a + b} = 1$, $\dfrac{a + b}{a - b}$ is the reciprocal of $\dfrac{a - b}{a + b}$, and vice versa.

Finding the Reciprocal of a Number

From the preceding definition and examples, it is clear that the reciprocal of an integer a is $1/a$, if $a \neq 0$.   The reciprocal of a fraction a/b is b/a, if $a \neq 0$ and $b \neq 0$.

EXAMPLE 8.   The reciprocal of x is $1/x$.

EXAMPLE 9.   The reciprocal of $\frac{3}{5}$ is $\frac{5}{3}$.

EXAMPLE 10.   The reciprocal of $xy/2$ is $2/xy$.

The Quotient of Two Fractions

As in arithmetic, if a/b and c/d are any two fractions, then

$$\frac{a}{b} \div \frac{c}{d} = \frac{a}{b} \cdot \frac{d}{c} = \frac{ad}{bc}$$

Expressed in words, the quotient of two fractions equals the dividend multiplied by the reciprocal of the divisor.   This is equivalent to the familiar rule in arithmetic: to divide one fraction by another, invert the divisor and multiply.   The student will be asked to justify this rule in Problem 48 of Exercise 30.

EXAMPLE 11. $\dfrac{3x}{2y} \div \dfrac{5x}{7y} = \dfrac{3\cancel{x}}{2\cancel{y}} \cdot \dfrac{7\cancel{y}}{5\cancel{x}} = \dfrac{21}{10}$

EXAMPLE 12.

$$\frac{x^2 + 3x - 10}{x^2 + 13x + 40} \div \frac{x^3 - 2x^2}{6x + 48} = \frac{\cancel{(x+5)}\cancel{(x-2)}}{\cancel{(x+5)}\cancel{(x+8)}} \cdot \frac{6\cancel{(x+8)}}{x^2\cancel{(x-2)}} = \frac{6}{x^2}$$

Exercise 30

MULTIPLICATION AND DIVISION OF FRACTIONS

Find each of the following products in lowest terms:

1. $\dfrac{2x}{3y} \cdot \dfrac{4y^2}{3x}$

2. $\dfrac{10an}{15n^2} \cdot \dfrac{12n}{5a^2}$

3. $\dfrac{5}{2x^2} \cdot \dfrac{6}{15x}$

4. $\dfrac{7}{a} \cdot \dfrac{4}{5}$

5. $\dfrac{a}{c} \cdot \dfrac{b}{3a^2}$

6. $\dfrac{a}{c} \cdot \dfrac{b}{4a}$

7. $\dfrac{25x^2}{6a^2x} \cdot \dfrac{4ax^2}{5a^2x^3}$

8. $\dfrac{3ax}{2ay} \cdot \dfrac{4y^2}{2ax}$

9. $\dfrac{13x^2}{7y^2} \cdot \dfrac{49y^3}{52x}$

10. $\dfrac{5ax}{2n} \cdot \dfrac{6nx}{15a^2}$

11. $\dfrac{x^2 - y^2}{x} \cdot \dfrac{x}{x - y}$

12. $\dfrac{n^2 - 4}{3n^2} \cdot \dfrac{n}{n - 2}$

13. $\dfrac{3a + b}{ab} \cdot \dfrac{a^2b^2}{3a - b}$

14. $\dfrac{x + 2}{x - 2} \cdot \dfrac{x^2 - 4x + 4}{x^2 - 4}$

15. $\dfrac{x^2 - 4}{x + 3} \cdot \dfrac{2x + 6}{4x - 8}$

16. $\dfrac{x + 5}{x^2 - 9} \cdot \dfrac{2x - 6}{2x + 10}$

17. $\dfrac{5a + 5b}{ac + bc} \cdot \dfrac{c^2a - c^2b}{a^2 - b^2}$ **18.** $\dfrac{3x^2 + 6}{5x^3} \cdot \dfrac{10x}{x^2 - 4}$

19. $\dfrac{x^2 - x - 6}{x^2 - 4} \cdot \dfrac{x + 2}{x - 3}$ **20.** $\dfrac{4a^2 - 100}{2a^2 - 32} \cdot \dfrac{a^2 + 8a + 16}{2a - 10}$

21. $\dfrac{x^2 + 3x}{x^2 - 9} \cdot \dfrac{3x^2 + 12x}{x^2 - 16} \cdot \dfrac{x^2 - 7x + 12}{3}$

Write the reciprocal of each of the following expressions. Verify that the product of each number by its reciprocal is equal to 1.

22. $\dfrac{1}{3}$ **23.** $-\dfrac{1}{5}$ **24.** -2

25. $\dfrac{3}{5}$ **26.** $\dfrac{3}{5x}$ **27.** $\dfrac{x}{2}$

28. $\dfrac{x + 2}{3}$ **29.** $\dfrac{2}{x - 2}$ **30.** $\dfrac{x^2 - a}{x^2 - 4}$

31. $\dfrac{a + b}{a + c}$ **32.** $\dfrac{a + 1}{b + 1}$ **33.** $\dfrac{a^2 - 2a + 8}{a^2 - 1}$

Express each of the following quotients in lowest terms:

34. $\dfrac{3ab}{x} \div \dfrac{a}{4x}$ **35.** $\dfrac{a^2}{n^2} \div \dfrac{2a}{n}$

36. $\dfrac{3xy}{2z} \div \dfrac{4x^2y}{9z}$ **37.** $\dfrac{3x}{2y} \div \dfrac{1}{y^2}$

38. $\dfrac{2a^2b}{3c} \div 4ab$ **39.** $\dfrac{2xy}{7} \div 4y^2$

40. $5x \div \dfrac{25x^2}{3}$ **41.** $(a + b) \div \dfrac{a^2 - b^2}{2}$

42. $\dfrac{15}{4x^2} \div \dfrac{5}{2x^6}$ **43.** $\dfrac{a + b}{a - b} \div \dfrac{2a + 4b}{2b - 2a}$

44. $\dfrac{x^2 - 1}{x^2 - 16} \div \dfrac{x - 1}{x^2 - 7x + 12}$ **45.** $\dfrac{x^2 - 3x + 2}{x^2 - 4} \div \dfrac{x^2 - 2x + 1}{2x - 4}$

46. $\dfrac{2x^2 - 18}{3x + 4} \div \dfrac{3x + 9}{9x^2 + 24x + 16}$ **47.** $\dfrac{x^2 - 9}{3x + 4} \div \dfrac{x + 3}{16 - 9x^2}$

48. Justify the rule for dividing fractions by writing

$$\frac{a}{b} \div \frac{c}{d} \text{ as } \frac{\dfrac{a}{b}}{\dfrac{c}{d}}$$

and then applying the fundamental principle of fractions.

31 Addition and Subtraction of Fractions

Lowest Common Multiple

A common multiple of two or more algebraic expressions is an expression which is exactly divisible by each of the given expressions.

EXAMPLE 1. $4xy^2$, $8x^2y^2$, $12xy^2$, etc., are each a common multiple of the expressions 2, x, and y^2.

The *lowest common multiple* (L.C.M.) is that multiple which is of lowest degree and which has the smallest numerical coefficient.

EXAMPLE 2. The L.C.M. of 2, x, and y^2 is $2xy^2$.

Finding the Lowest Common Multiple

When the expressions are simple, the lowest common multiple may be found by inspection.

EXAMPLE 3. The L.C.M. of $3x^2$, $5xy$, and $6y^2$ is $30x^2y^2$, since it is the expression of lowest degree and with smallest coefficient, which is divisible by each of the given expressions.

When the lowest common multiple is not evident by inspection, it may be found by the following rule:

1. Find the prime factors of each expression.

2. The lowest common multiple is the product of each prime factor raised to the highest power of occurrence in any one expression.

EXAMPLE 4. Find the L.C.M. of $9x^2$, $15xy$, $24xy^2$, and $35x^2y^2$.

$$9x^2 = 3^2x^2$$
$$15xy = 3 \cdot 5 \cdot x \cdot y$$
$$24xy^2 = 2^3 \cdot 3 \cdot x \cdot y^2$$
$$35x^2y^2 = 5 \cdot 7 \cdot x^2 \cdot y^2$$
$$\text{L.C.M.} = 2^3 \cdot 3^2 \cdot 5 \cdot 7 \cdot x^2 \cdot y^2 = 2520x^2y^2$$

EXAMPLE 5. Find the L.C.M. of $x^3 - x$, $x^3 - x^2 - 2x$, and $x^2 - 2x + 1$.

$$x^3 - x = x(x - 1)(x + 1)$$
$$x^3 - x^2 - 2x = x(x - 2)(x + 1)$$
$$x^2 - 2x + 1 = (x - 1)^2$$
$$\text{L.C.M.} = x(x - 1)^2(x + 1)(x - 2)$$

Note. In a problem such as Example 5, it is convenient to express the L.C.M. in factored form, rather than to multiply the factors as in Example 4.

Addition and Subtraction of Fractions

As in arithmetic, algebraic fractions may be added or subtracted provided they have like denominators. We therefore have the following procedure for adding or subtracting fractions:

1. Find the lowest common multiple of the denominators of the fractions. This is called the lowest common denominator (L.C.D.).

2. Change each fraction to an equivalent fraction having the lowest common denominator for its denominator. To avoid errors, the actual multiplication is not carried out until all of the numerators are written over the common denominator, as in Examples 6 and 7.

3. The sum or difference of the fractions will be a fraction whose numerator is the sum or difference of the numerators and whose denominator is the lowest common denominator.

EXAMPLE 6. The sum of $\dfrac{3x - 2}{4x} + \dfrac{2y - 1}{5y}$

$$= \frac{5y(3x - 2)}{20xy} + \frac{4x(2y - 1)}{20xy}$$

$$= \frac{5y(3x - 2) + 4x(2y - 1)}{20xy}$$

$$= \frac{23xy - 10y - 4x}{20xy}$$

EXAMPLE 7. The sum of $\dfrac{3}{x^2 - 4} - \dfrac{2}{x^2 - 3x + 2}$

$$= \frac{3}{(x - 2)(x + 2)} - \frac{2}{(x - 2)(x - 1)}$$

$$= \frac{3(x - 1)}{(x - 2)(x + 2)(x - 1)} - \frac{2(x + 2)}{(x - 2)(x - 1)(x + 2)}$$

$$= \frac{3(x - 1) - 2(x + 2)}{(x - 2)(x + 2)(x - 1)} = \frac{x - 7}{(x - 2)(x + 2)(x - 1)}$$

Complex Fractions

A *complex fraction* is a fraction which contains one or more fractions in either its numerator or its denominator, or in both.

EXAMPLE 8. $\dfrac{x + \frac{1}{2}}{x}$ is a complex fraction.

EXAMPLE 9. $\dfrac{x - \frac{1}{3}}{\dfrac{x}{2} - \dfrac{1}{x}}$ is a complex fraction.

Simplifying a Complex Fraction

To simplify a complex fraction, we may use the fundamental principle of fractions, multiplying both numerator and denominator of the complex fraction by the lowest common denominator of the fractions in the numerator and denominator.

EXAMPLE 10. $\dfrac{4 - \frac{1}{9}}{2 - \frac{2}{3}} = \dfrac{9(4 - \frac{1}{9})}{9(2 - \frac{2}{3})} = \dfrac{36 - 1}{18 - 6} = \dfrac{35}{12}$

EXAMPLE 11. $\dfrac{x/y}{y/x - x/y} = \dfrac{xy\,(x/y)}{xy\,(y/x - x/y)} = \dfrac{x^2}{y^2 - x^2}$

EXAMPLE 12. $\dfrac{1 - 1/x}{1 - 1/x^2} = \dfrac{x^2(1 - 1/x)}{x^2(1 - 1/x^2)} = \dfrac{x^2 - x}{x^2 - 1} = \dfrac{x}{x + 1}$

Exercise 31

ADDITION AND SUBTRACTION OF FRACTIONS

Find the L.C.M. (in factored form) of each of the following sets of expressions:

1. $18, 24, 36$ 2. ax, a^2x^2, ax^2
3. $3ax, 5ax^2, 7a$ 4. $4x^2, 6nx^2, 9n^2x$
5. $2\pi r, 4\pi r^2, \pi r^2 h$ 6. $3x^2, 15xy, 35x^2y$
7. $3ax, 9ay, 12az$ 8. $6xy, 12xz, 15yz$
9. $36xy^2, 42x^2y, 63xy^2$ 10. $5a^2b, 7ab^2, 35abc$
11. $4x, x^2 - 2x$ 12. $6xy, 3xy^2 - 3x^2y$
13. $2x + 4y, 4x + 8y$ 14. $a^2 - ax, 3ab - 3ax$
15. $4x, 2x^2 - 2x, 2x^2 - 4x$ 16. $x^2 - 9, x^2 + 4x + 3$
17. $x^2 - 2x + 1, x^2 - 1, 2x$ 18. $x^2 - 2x, x^2 + 2x, 2x^2 - 8$
19. $x - 2y, xy - 2y^2, x^2 - 4y^2$ 20. $x - 3, x^2 + 2x - 15, x + 5$

Perform the indicated operations and simplify the results:

21. $\dfrac{5x}{5} + \dfrac{2x}{5}$

22. $\dfrac{3x}{2} + \dfrac{x}{3} + \dfrac{4x}{15}$

23. $\dfrac{3x}{4} + \dfrac{2x^2}{5}$

24. $\dfrac{x+1}{3} + \dfrac{x-1}{3}$

25. $\dfrac{2x-3}{3} - \dfrac{x+2}{4}$

26. $\dfrac{2x-3}{2} - \dfrac{x-2}{5}$

27. $\dfrac{x-3}{5} - \dfrac{x+3}{6}$

28. $\dfrac{x+7}{2} - \dfrac{3x-2}{4}$

29. $\dfrac{7}{5x} + \dfrac{12}{5x}$

30. $\dfrac{3}{2x} - \dfrac{2}{3x}$

31. $\dfrac{2x}{3} + \dfrac{4x^2}{5}$

32. $\dfrac{3x}{2y} - \dfrac{2x}{3y} + \dfrac{4}{y}$

33. $\dfrac{2x}{4} + \dfrac{3}{5x} + \dfrac{7}{5x^2}$

34. $\dfrac{4}{5y^2} - \dfrac{2}{5y}$

35. $\dfrac{2}{3y} - \dfrac{4}{2x}$

36. $\dfrac{2}{x-1} + \dfrac{3}{x+1}$

37. $2 + \dfrac{1}{x-1}$

38. $\dfrac{2x}{x-3} + \dfrac{4x}{x-2}$

39. $\dfrac{2a}{a-1} - \dfrac{3}{a} + \dfrac{2a}{a+1}$

40. $\dfrac{1}{x-2} - \dfrac{2}{x^2-4} + \dfrac{3}{x+2}$

41. $\dfrac{x-1}{x^2-x-6} + \dfrac{x}{x-3}$

42. $\dfrac{2x}{x^2-4} + \dfrac{3x}{x^2-2x+4}$

43. $\dfrac{2x}{x^2-y^2} + \dfrac{3x}{x^2-xy-2y^2} + \dfrac{3xy}{x^2-4y^2}$

44. $3x - 2 - \dfrac{2}{x} + \dfrac{x-1}{x+1}$

45. $\dfrac{x-2}{x} - 2 + \dfrac{x-2}{x-1}$

46. $\dfrac{2x+4}{x^2+3x-10} - \dfrac{2}{2x} + \dfrac{3x-2}{x^2+5x}$

47. $\dfrac{x}{x^2+3x+2} + \dfrac{2x}{x^2-1} + \dfrac{2}{x^2+x-1}$

48. $\dfrac{1}{x} - \dfrac{2x-6}{x^2-6x} + \dfrac{1}{x-6}$

49. $\dfrac{4x^2-14x-32}{x^2-4x} + \dfrac{x-7}{4-x}$

50. $\dfrac{3x-7}{x^2-5x+4} - \dfrac{x+4}{x-1} + \dfrac{x}{x-4}$

51. $\dfrac{x^2+3}{x^4-16} - \dfrac{2}{x^2-4} + \dfrac{3}{x^2+4}$

52. $\dfrac{x+3}{x^2+x} + \dfrac{2}{x} + \dfrac{x-5}{x^2+2x+1}$

53. $\dfrac{n+3}{n^2-8n+15} - \dfrac{n}{5-n} - \dfrac{3}{3-n}$

54. $\dfrac{x+4}{x^2-5x+6} - \dfrac{3}{3-x} - \dfrac{2}{x-2}$

55. $\dfrac{x^2+4}{3x-2} + 3x + 2$

56. $x^2 + y^2 - \dfrac{x^3+y^3}{x+y} - xy$

57. $\dfrac{x}{x-2} - \dfrac{x-2}{x+2} - \dfrac{8}{x^2-4}$

58. $\dfrac{n}{n+1} + \dfrac{2n^2}{n^2-1} - \dfrac{2n}{n-1}$

59. $\dfrac{1}{x} + \dfrac{2x-y}{x^2+xy}$

60. $\dfrac{1}{x-1} - \dfrac{1}{x-2} + \dfrac{2}{(x-1)(2-x)}$

61. $\dfrac{1+\dfrac{x}{2}}{1-\dfrac{x}{2}}$

62. $\dfrac{\dfrac{1+x}{x}}{\dfrac{1-x}{x}}$

63. $\dfrac{1-\dfrac{3}{a}}{1+\dfrac{4}{2a}}$

64. $\dfrac{x-\dfrac{1}{2}}{1-\dfrac{2}{x}}$

65. $\dfrac{\dfrac{1+x}{x}}{1-\dfrac{1}{x^2}}$

66. $\dfrac{1+\dfrac{2}{x-y}}{\dfrac{2}{x-y}}$

67. $\dfrac{1+\dfrac{2}{xy}}{\dfrac{x-y}{x}}$

68. $\dfrac{\dfrac{1}{x}+\dfrac{2}{x^2}-\dfrac{3}{x^3}}{\dfrac{1}{x^2}-\dfrac{2}{x^4}}$

69. $\dfrac{\dfrac{1}{x}+\dfrac{1}{y}+\dfrac{1}{xy}}{\dfrac{2}{x}-\dfrac{2}{y}}$

70. $\dfrac{\dfrac{2}{x+y}+1}{\dfrac{2}{x+y}-1}$

71. $\dfrac{\dfrac{x}{x+1}-\dfrac{4}{x+6}}{\dfrac{2x}{x^2+7x+6}}$

72. $\dfrac{5-\dfrac{3}{t+1}}{20+\dfrac{9}{t^2-1}}$

73. $\dfrac{\dfrac{x^2+2x-1}{x^2-25}-1}{5+\dfrac{3x+5}{x+5}}$

74. $\dfrac{2-\dfrac{x+1}{x-3}}{1-\dfrac{6x-3}{x^2-9}}$

32   Solution of Fractional Equations

Fractional Equations

If at least one of the terms of an equation is a fraction, the equation is called a fractional equation. (In the strictest sense, an equation is classified as fractional only if at least one of the denominators contains the unknown, but we shall use the more general definition first stated.) Examples of fractional equations are:

$$\frac{1 + x}{2} + \frac{2x}{7} = 5$$

$$\frac{4}{x} + \frac{2x}{x - 1} - \frac{3}{2x} = 4$$

Solving a Fractional Equation

To solve a fractional equation, we use the same principles of equality discussed in Chapter 4. As a preliminary step, however, it is necessary to clear the equation of fractions, that is, find an equivalent equation without fractions. Consider the equation

$$\frac{x}{2} + \frac{x - 1}{3} = 8$$

If we multiply each term of the equation by 6 (the least common denominator), we shall have

$$6\left(\frac{x}{2}\right) + 6\left(\frac{x - 1}{3}\right) = 6(8)$$
$$\text{or} \quad 3x + 2x - 2 = 48$$

which may be solved by the methods of Chapter 4. It is evident that since each denominator of a fraction is an exact divisor of the L.C.D., multiplying each term of any fractional equation by the L.C.D. will eliminate all fractions, as in the preceding example.

In an equation with numerical denominators, multiplying by a common denominator greater than the L.C.D. will not affect the final result. The preceding equation, for example, could have been cleared of fractions by multiplying by 12 or 18. In an equation whose denominators contain the unknown, however, unless the L.C.D., is used to clear the equation of fractions, *extraneous* roots (that is, apparent roots which do not satisfy the original equation) may be introduced.   In fact, the introduction of extraneous roots is always a possibility when an equation is multiplied by an expression containing the unknown, so that checking the solution is particularly important in fractional equations.   Examples 4 and 5 are equations in which extraneous roots are introduced.

EXAMPLE 1.   Solve the equation $x + \dfrac{x - 6}{6} = \dfrac{5}{2}$.

$$\frac{x}{1} + \frac{x - 6}{6} = \frac{5}{2} \tag{1}$$

The L.C.D. = 6.

$$6(x) + 6\left(\frac{x - 6}{6}\right) = 6\left(\frac{5}{2}\right) \tag{2}$$

$$6x + x - 6 = 15 \tag{3}$$

$$x = 3 \tag{4}$$

Check: Substituting $x = 3$ in the given equation,

$$3 + \frac{3 - 6}{6} = \frac{5}{2} \tag{5}$$

$$\frac{5}{2} = \frac{5}{2}$$

With a little practice, the indicated multiplication in Equation 2 may be done mentally, so that Equation 3 may be written as soon as the L.C.D. is determined.

When the denominators are expressions containing the unknown, it is convenient to write the equation with the denominators in factored form in order to make the common denominator evident.   This also simplifies the multiplication of each term by the L.C.D., since like factors can be divided out at once.

EXAMPLE 2.   Solve the equation $\dfrac{2x}{x^2 - 1} - \dfrac{x + 1}{x - 1} = \dfrac{1 - x}{x + 1}$.

$$\frac{2x}{(x - 1)(x + 1)} - \frac{x + 1}{x - 1} = \frac{1 - x}{x + 1} \tag{1}$$

The L.C.D. is $(x - 1)(x + 1)$

$$2x - (x + 1)(x + 1) = (x - 1)(1 - x) \tag{2}$$

$$2x - x^2 - 2x - 1 = -x^2 + 2x - 1 \tag{3}$$

$$x = 0 \tag{4}$$

If $x = 0$ is substituted in the original equation both members of the equation will equal 1, so that $x = 0$ is a solution of the equation.

EXAMPLE 3. Solve the equation $\dfrac{x + 2}{x + 3} + \dfrac{x + 3}{x + 2} = \dfrac{11x + 28}{x^2 + 5x + 6}$.

$$\frac{x + 2}{x + 3} + \frac{x + 3}{x + 2} = \frac{11x + 28}{(x + 3)(x + 2)} \tag{1}$$

The L.C.D. $= (x + 3)(x + 2)$

$$(x + 2)(x + 2) + (x + 3)(x + 3) = 11x + 28 \tag{2}$$

$$2x^2 - x - 15 = 0 \tag{3}$$

$$(2x + 5)(x - 3) = 0 \tag{4}$$

$$x = 3, x = -\tfrac{5}{2} \tag{5}$$

On substituting $x = 3$ and $x = -\tfrac{5}{2}$ in the original equation, it will be found that both roots satisfy the equation.

EXAMPLE 4. Solve the equation $\dfrac{1}{x + 1} - \dfrac{2x + 3}{x + 1} = 2$.

$$\frac{1}{x + 1} - \frac{2x + 3}{x + 1} = \frac{2}{1} \tag{1}$$

The L.C.D. $= x + 1$

$$1 - (2x + 3) = 2(x + 1) \tag{2}$$

$$x = -1 \tag{3}$$

Check: $\qquad \dfrac{1}{-1 + 1} - \dfrac{2(-1) + 3}{-1 + 1} = 2$

Since each denominator equals 0, $x = -1$ cannot be considered as a root of the equation and therefore the equation has no solution.

EXAMPLE 5. Solve the equation $\dfrac{x - 1}{x^2 - 5x + 6} + \dfrac{x - 2}{x^2 - 4x + 3} + \dfrac{1}{x - 2} = 0$.

$$\frac{x - 1}{(x - 3)(x - 2)} + \frac{x - 2}{(x - 1)(x - 3)} + \frac{1}{(x - 2)} = 0 \tag{1}$$

The L.C.D. $= (x - 3)(x - 2)(x - 1)$

$$(x - 1)(x - 1) + (x - 2)(x - 2) + (x - 3)(x - 1) = 0 \tag{2}$$

$$3x^2 - 10x + 8 = 0 \tag{3}$$

$$x = 2 \text{ and } x = \tfrac{4}{3} \tag{4}$$

The root $x = 2$ is extraneous, since two of the denominators equal zero when $x = 2$. The value $x = \tfrac{4}{3}$ satisfies the original equation, and is therefore the only solution.

Transformation of Formulas

An important application of fractional equations occurs when it is necessary to change the form of certain formulas.   The formula $F = \frac{9}{5}C + 32$ is used to convert degrees centigrade to degrees Fahrenheit.   If we wish to change the formula so that we may convert degrees Fahrenheit to degrees centigrade, we proceed as follows:

$$F = \tfrac{9}{5}C + 32 \tag{1}$$

The L.C.D. is 5 $\tag{2}$

$$5F = 9C + 160 \tag{3}$$

$$C = \frac{5F - 160}{9} \tag{4}$$

EXAMPLE 6.   In the formula for parallel resistances in an electrical circuit,

$$\frac{1}{R} = \frac{1}{r_1} + \frac{1}{r_2}$$

express r_1 in terms of R and r_2.

$$\frac{1}{R} = \frac{1}{r_1} + \frac{1}{r_2} \tag{1}$$

The L.C.D. $= Rr_1r_2$ $\tag{2}$

$$r_1r_2 = Rr_2 + Rr_1 \tag{3}$$

$$r_1[r_2 - R] = Rr_2 \tag{4}$$

$$r_1 = \frac{Rr_2}{r_2 - R} \tag{5}$$

Exercise 32

SOLUTION OF FRACTIONAL EQUATIONS

Solve each of the following equations:

1. $\dfrac{x}{2} + \dfrac{x}{4} = 6$

2. $\dfrac{x}{3} + \dfrac{5x}{9} = 8$

3. $\dfrac{x}{2} + \dfrac{x}{3} + \dfrac{x}{4} = 26$

4. $\dfrac{2x}{5} + \dfrac{3x}{2} = 19$

5. $\dfrac{x + 4}{2} - \dfrac{x - 3}{5} = 5$

6. $\dfrac{3x + 1}{4} + \dfrac{5x - 1}{6} = 8$

7. $\dfrac{x + 9}{2} - \dfrac{5x + 3}{6} = 2$

8. $\dfrac{x - 7}{2} - \dfrac{x - 1}{8} = \dfrac{3}{4}$

9. $\dfrac{6x}{5} - \dfrac{2x + 12}{3} = \dfrac{-44}{15}$

10. $\dfrac{x + 3}{5} + \dfrac{x - 3}{2} = 4$

11. $\dfrac{3x - 1}{2} + \dfrac{4x}{7} = \dfrac{2x + 11}{14}$

12. $\dfrac{7x - 2}{3} - \dfrac{4x - 1}{5} = \dfrac{3 - 3x}{5}$

13. $\dfrac{x + 3}{2} + \dfrac{x + 5}{6} = \dfrac{x + 1}{9}$

14. $\dfrac{2x - 1}{5} + \dfrac{3x - 2}{7} = \dfrac{7x + 7}{14}$

15. $\dfrac{x + 1}{10} - \dfrac{x + 3}{2} = \dfrac{x - 24}{3}$

16. $\dfrac{3x - 2}{2} + \dfrac{12x + 3}{10} = \dfrac{6x - 1}{5}$

Solve each of the following equations:

17. $\dfrac{1}{x} + \dfrac{2}{x} + \dfrac{3}{x} = 6$

18. $\dfrac{1}{2x} - \dfrac{1}{4x} = \dfrac{1}{6} + \dfrac{1}{3x}$

19. $\dfrac{3}{x} - \dfrac{2}{x} = -5$

20. $\dfrac{6}{x} - \dfrac{x + 12}{3x} = \dfrac{4}{3}$

21. $\dfrac{3}{4x} + \dfrac{6}{5x} = \dfrac{13}{20}$

22. $\dfrac{6}{x} - \dfrac{4}{5x} = 13$

23. $\dfrac{5x + 6}{2x} - \dfrac{7x - 1}{5x} = -\dfrac{1}{2}$

24. $\dfrac{6x - 5}{2} + \dfrac{25}{18x} = \dfrac{9x + 5}{3}$

25. $\dfrac{x - 8}{2x} + \dfrac{x - 2}{x} = \dfrac{5}{2}$

26. $\dfrac{x}{5 - x} = \dfrac{2x}{4x - 5}$

27. $\dfrac{x + 7}{3 - x} + \dfrac{3x}{2} = x$

28. $\dfrac{x + 29}{x + 9} + 1 = \dfrac{2x + 5}{x - 2}$

29. $\dfrac{x + 7}{3 - x} = \dfrac{24x - 3}{x + 3}$

30. $\dfrac{y + 6}{3 - y} = \dfrac{2y + 5}{7y - 5}$

Solve each of the following equations:

31. $\dfrac{2}{x - 1} - \dfrac{3}{x + 3} = \dfrac{6}{x^2 + 2x - 3}$

32. $\dfrac{5}{t + 1} + \dfrac{10}{t^2 - t - 2} = \dfrac{4}{t - 2}$

33. $\dfrac{3}{x - 2} - \dfrac{5}{x + 4} = \dfrac{10}{x^2 + 2x - 8}$

34. $\dfrac{9}{x - 4} - \dfrac{5}{x + 1} = \dfrac{27}{x^2 - 3x - 4}$

35. $\dfrac{x - 12}{x^2 - 10x + 25} - \dfrac{1}{x} = \dfrac{3}{x^2 - 5x}$

36. $\dfrac{x}{x - 8} - \dfrac{5}{x + 8} = \dfrac{x^2 + 64}{x^2 - 64}$

37. $8 - x + \dfrac{x}{x - 9} = \dfrac{7}{9 - x}$

38. $\dfrac{x - 3}{x + 3} - \dfrac{x + 2}{x - 2} = 0$

39. $\dfrac{2x - 1}{4x^2 - 10x + 6} + \dfrac{2x + 1}{2x - 3} = 1$

40. $\dfrac{x - 4}{x + 3} + \dfrac{x + 1}{x + 2} = \dfrac{2x^2}{x^2 + 5x + 6}$

41. $\dfrac{2x+1}{5x+3} + \dfrac{1}{12} = \dfrac{x+1}{x-1}$ **42.** $\dfrac{x+3}{x} - \dfrac{x+4}{x+5} = \dfrac{15}{x^2+5x}$

43. $\dfrac{2x+1}{2x-3} - \dfrac{x-4}{2x+3} = 1 - \dfrac{7x}{9-4x^2}$

44. $\dfrac{1}{x^2-3x+2} = \dfrac{2}{4-x^2} + \dfrac{1}{1-x}$

Solve each of the following formulas for the indicated letter:

45. $E = IR$ (R) **46.** $E = \tfrac{1}{2}mv^2$ (m)

47. $S = -\tfrac{1}{2}gt^2 + V_0 t$ (V_0) **48.** $\dfrac{1}{F} = \dfrac{1}{f_1} + \dfrac{1}{f_2}$ (f_2)

49. $F = \dfrac{m_1 m_2}{d^2}$ (m_1) **50.** $P = I^2 R$ (R)

51. $S = \dfrac{rl-a}{r-1}$ (a) **52.** $I = \dfrac{E}{r + \dfrac{R}{n}}$ (n)

53. $R = \dfrac{V_1 V_2}{V_1 + V_2}$ (V_2) **54.** $R = \dfrac{1}{2\pi f c}$ (f)

55. $W = \dfrac{2PR}{R-r}$ (r) **56.** $s = \dfrac{n}{2}(a+l)$ (l)

57. $P = \dfrac{s}{1+in}$ (n) **58.** $A = \tfrac{1}{2}h(b_1 + b_2)$ (b_1)

59. $C = \dfrac{en}{R+nr}$ (R) **60.** $V = \dfrac{h(b_1 + b_2 + 4M)}{6}$ (M)

33   Applications of Fractional Equations

We now consider problems which, when expressed in symbolic form, result in fractional equations.   There are many types of problems of this kind, but we shall illustrate just a few of the more typical ones.

EXAMPLE 1.   A package of grass seed contains a mixture of 1 pound of perennial rye and 4 pounds of marion blue.   How much perennial rye should be added to have a mixture which is one-half perennial rye?   Let x = number of pounds of perennial rye to be added.   Then $x + 1$ is the new amount of perennial rye and $x + 5$ is the total amount of the mixture.   Therefore

$$\frac{x+1}{x+5} = \frac{1}{2}$$
$$x = 3 \text{ pounds of perennial rye}$$

When 3 pounds of rye have been added, there will be 4 pounds of rye and 8 pounds of the mixture.   Therefore the mixture is one-half rye.

EXAMPLE 2.   One printing press can print a certain number of papers in 4 hours; another smaller press can print the same number in 6 hours. In how many hours can both presses working together print the papers?

Let x = total time for both presses.   Then $x/4$ is the fractional part of the work done by the first press in x hours.   Similarly, $x/6$ is the fractional part of the work done by the second press in x hours.   Therefore

$$\frac{x}{4} + \frac{x}{6} = 1$$

$x = 2\frac{2}{5}$ hours, or 2 hours and 24 minutes

In $2\frac{2}{5}$ hours, the first press does $2\frac{2}{5}/4$ or $\frac{3}{5}$ of the work, and the second press does $2\frac{2}{5}/6$ or $\frac{2}{5}$ of the work.   Since $\frac{3}{5} + \frac{2}{5} = 1$, the job is finished in $2\frac{2}{5}$ hours.

EXAMPLE 3.   An automobile and a truck leave at the same time for a point 300 miles away.   The truck averages 15 miles per hour less than the automobile, and arrives 1 hour and 40 minutes later than the car. What is the rate of each?

Both the rates of speed and the time are unknown, but using the relation $t = d/r$, we may express the difference in time in terms of the unknown rates.   Let x = rate of truck.   Then $x + 15$ = rate of automobile.   Therefore

$$\frac{300}{x} - \frac{300}{x + 15} = \frac{5}{3}$$

The solution of the resulting quadratic equation is $x = 45$ and $x = -60$. Clearly, $x = -60$ must be discarded as an answer, so that $x = 45$ miles per hour, the rate of the truck; $x + 15 = 60$ miles per hour, the rate of the automobile.   The time of the truck is $300/45$, or 6 hours and 40 minutes.   The time of the automobile is $300/60$, or 5 hours.   The difference in time is 1 hour and 40 minutes.

Exercise 33

APPLICATIONS OF FRACTIONAL EQUATIONS

1. An automobile radiator contains 12 quarts of water and 2 quarts of antifreeze.   How much antifreeze should be added to produce a mixture which is one-sixth antifreeze?

2. A girl's age is three-fifths of what it will be 10 years from now. How old is she now?

3. One-eighth of a man's age 5 years ago equals one-tenth of his age 5 years from now. How old is he?

4. A is twice as old as B. Four years from now the ratio of their ages will be $\frac{5}{3}$. What are their ages now?

5. A concrete wall has one-fourth of its total height in the ground and one-third of its total height in water. If 10 feet of the wall rise above the water, what is the total height of the wall?

6. Two-thirteenths of the supplement of an angle plus one-eighth of the complement of the same angle equals one-half the angle. Find the number of degrees in the angle.

7. Three-sevenths of a number plus the number itself equals 80. What is the number?

8. The sum of a number and its reciprocal is $2\frac{9}{10}$. Find the number.

9. A certain number is added to both numerator and denominator of the fraction $\frac{17}{21}$. The resulting fraction equals $\frac{1}{2}$. What is the number?

10. What single number can be added to both terms of each of the fractions $\frac{7}{8}$ and $\frac{27}{31}$ to make then equal?

11. There are three consecutive even integers. One-sixth of the first plus one-fourth of the second plus one-half of the third equals 19. Find the three integers.

12. The denominator of a fraction is 7 more than the numerator. If 1 is added to the numerator and 4 is subtracted from the denominator, the resulting fraction equals $\frac{1}{2}$. What is the original fraction?

13. The numerator of a fraction is 10 less than the denominator. If 4 is subtracted from the numerator and added to the denominator, the resulting fraction equals $\frac{1}{8}$. What is the fraction?

14. When 63 is divided by a certain number, the quotient is 5, and the remainder is 3 less than the divisor. What is the divisor?

15. The sum of a number and its reciprocal is 3 less than the number itself. Find the number.

16. The sum of the reciprocals of two consecutive even integers is $\frac{9}{40}$. Find the integers.

17. The numerator of a positive fraction is 5 less than the denominator. When 3 is added to both numerator and denominator, the original fraction is increased by one-twelfth. What is the fraction?

18. The denominator of a certain positive fraction is 1 more than the numerator. The fraction plus its reciprocal equals $2\frac{1}{6}$. What is the fraction?

19. Separate 65 into two parts so that one-third of the larger part is 5 more than one-half of the smaller part.

20. Teacher A can mark a set of examination papers in 3 hours, and teacher B can mark the same set in 2 hours. How long will it take to mark the set if both teachers work together?

21. One crew can unload a ship is 5 hours. Another crew can unload the same ship in $4\frac{1}{2}$ hours. How many hours will it take if both crews work together?

22. A tank can be filled by one pipe in 3 hours and another in 4 hours. It can be emptied in 5 hours by another pipe. If all pipes are open, how long will it take to fill the tank?

23. John can do a job in 8 hours. After working 3 hours, Joe joins him, and they finish the job in two more hours. How long would it have taken Joe to do the job alone?

24. Bob can paint a garage in 9 hours. Bob and Sam working together can paint it in 5 hours. How long would it take Sam if he worked alone?

25. A factory produced 1000 radios in 4 days. A second factory was built, and the combined production rose to 1000 radios in $2\frac{1}{2}$ days. In how many days could the second factory produce 1000 radios?

26. A tank can be filled by one pipe in 1 hour less time than by another. With both pipes open, it takes $2\frac{1}{2}$ hours to fill the tank. In how many hours can each pipe alone fill the tank?

27. An accountant has 10 days in which to finish an income tax report. Without a calculator, the report would take 16 days, and with a calculator, which rents for $10.00 a day, only 8 days would be required. What is the least amount he must pay for the use of the calculator in order to complete the job on time?

28. The rate of a river is 4 miles per hour. If a boat can go 8 miles downstream in the same time it takes to go 4 miles upstream, what is the rate of the boat?

29. A ferry boat with a rate of 9 miles per hour in still water takes $\frac{1}{2}$ hour to travel 2 miles downstream and back. What is the rate of the current?

30. A train makes a run of 108 miles at a certain rate. Increasing the rate by 6 miles per hour saves 12 minutes. What was the original rate?

31. A train travels 150 miles in 1 hour and 15 minutes less time than a car.   If the train traveled 20 miles per hour faster than the car, what is the rate of each?

32. A passenger train has twice the speed of a freight train.   If it takes 15 hours more for the freight train to travel 700 miles than it takes the passenger train to travel 350 miles, find the rate of each.

33. A car made a round trip in 5 hours.   If the total distance was 80 miles and the rate going averaged 30 miles per hour, what was the average rate returning?

34. The rates of two trains differ by 10 miles per hour.   The faster train travels 800 miles in 4 hours less time than the slower train.   Find the rate of each.

35. A train running at 45 miles per hour requires 2 hours longer between stations than another train running at 60 miles per hour.   What is the distance between stations?

36. A car traveled on a 270 mile toll road a distance of 200 miles before stopping for 2 hours.   The rest of the trip, including the 2 hour stop, takes the same time as the first part.   At what rate did the car travel?

37. A man can row 8 miles downstream and back in 6 hours.   If the rate of the current is 2 miles per hour, what is the rate of the boat in still water?

CHAPTER 7

Functional Relationships

The idea of relation is used in mathematics, as well as other fields, to describe a correspondence between the elements of two sets. For example, the relation "brother" can be used to describe a correspondence between certain individuals. We can say that A is the brother of B, or that A is related to B by the brother relation. The relation "is the capital of" sets up a correspondence between certain cities and their respective states or countries. The teacher-student relation describes the correspondence between the members of a set of teachers on the one hand, and the members of a set of students on the other hand.

In some cases the correspondence described by the relation may be one-to-many, as, for example, the relation of a teacher to students in a classroom. In other cases there may be a single element in a set corresponding to each element of another set, as in the marriage relation in a monogamous society. This last example is an informal illustration of a particular class of relations called *functions*. More explicitly, a function is a relation which establishes a correspondence of one element of a second set with each given element of a first set.

The world about us contains numerous important examples of the function relation. We note that the length of a metal rod changes with its temperature, the height to which a projectile rises in free flight depends on its initial upward velocity, and the radioactivity of a sample of radium changes with respect to time. In each case there is a correspondence between elements of two sets—the one set represented by values of a given variable and the other set represented by values of a related variable.

160

The mathematical description of a function may take the form of an equation, and this is usually the goal.   However, a function is not necessarily expressible in terms of an equation; at times a verbal statement, a table of values, or a graph may be used to describe the relation.   The mathematical description of a function is used to provide useful information about the basic pattern of behavior of the related variable and makes possible the prediction of results for selected values of the variables.

Functions are important in such areas as the physical, biological, and social sciences when studies proceed beyond a rudimentary level.   A significant application in these fields, however, requires considerable knowledge of both mathematics and the particular science in which the variables occur.   Therefore our study of functions will be confined to some elementary aspects of the language, symbolism, and methods of representation of these relations.

Definition of Function

Having discussed the concept of function in an informal way, we now consider a more rigorous and more useful mathematical interpretation. A variable, it will be recalled, is a symbol which denotes any one of a given set of numbers.   Consider now the equality

$$y = 5x + 4$$

where x is a variable limited to the set of real numbers.   The following conclusions can be readily verified:

1. The expression $5x + 4$ is also a variable since it changes as x changes.
2. The literal quantity y is a variable since it is equal to $5x + 4$.
3. For each value of x, there exists a corresponding value of y.
4. The value of y depends on the value of x.

This example illustrates a common form of the mathematical statement of a function.

By generalizing the significant features of this relationship, we now define a function as follows:

A function is a system consisting of two sets and a rule of correspondence which assigns to each value of a variable x in one set, not more than one corresponding value of a variable y in the other set.

In the illustration above, the first set consists of all admissible values of x; it is called the *domain* of the function.   The second set consists of all corresponding values of the expression $5x + 4$; it is called the *range* of the function. The values of $5x + 4$ are designated more briefly by the symbol y. The rule of correspondence is that for each value of x, y is obtained by adding 4 to the product of 5 and x. For example, if $x = 4$, the rule of correspondence determines that $y = 24$, and if $x = -3$, the corresponding value of y is -11. Since, in a similar manner each value of x determines a corresponding value of y, we say that y depends on x. For this reason, y is called the *dependent variable*, and x is called the *independent variable*.

It is clear that y is the value of the function at each corresponding value of x. In common practice, however, we often abbreviate this to the expression that y *is a function of* x.

EXAMPLE 1. The expression $3x^2$ is a function of x since each value of x determines a corresponding value of $3x^2$. The independent variable is x; the dependent variable is the expression $3x^2$.

EXAMPLE 2. The circumference of a circle is a function of its radius because for each value of the radius, r, the circumference, C, is determined by the relationship $C = 2\pi r$. The domain of r is the set of all positive real numbers. The independent variable is r; the dependent variable is C.

EXAMPLE 3. The temperature, T, of the air at a given weather station is a function which assigns to each time, t, a corresponding temperature. The independent variable is t whose domain is the set of time measurements. The dependent variable is T whose range is the corresponding set of temperatures. Unlike Example 2, however, this function cannot generally be expressed in the form of an equation.

Functional Notation

It is frequently necessary to discuss properties of a function without specifying its particular form. If y is a function of x, this fact may be denoted without assigning the particular form of the relationship by using symbols such as

$$y = f(x), \qquad y = F(x), \qquad y = y(x)$$

The symbol $y = f(x)$ is read "y is a function of x" or "y equals f of x." It is important to note that this symbol does not mean f multiplied by x.

EXAMPLE 4. The volume of a sphere is a function of its radius. We can indicate this by the statement $V = f(r)$. The surface area of a sphere is also a function of its radius. Since it is a different function, we can use the notation $S = g(r)$.

An important feature of this notation is that if $f(x)$ represents a given function of x , then $f(a)$ represents the value of the function when x assumes the value a .

EXAMPLE 5. The symbol $f(x)$ is defined to represent the function $5x^2 + 3$. What is the value of $f(2)$? The value of $5x^2 + 3$ when x is replaced by 2 is 23. Therefore $f(2) = 23$.

EXAMPLE 6. If $f(x) = x^2 + x - 9$, find $f(-6)$.

$$f(-6) = (-6)^2 + (-6) - 9 = 21$$

EXAMPLE 7. Show that $f(a) = f(-a)$ if $f(x) = x^2$.

$$f(x) = x^2$$
$$f(a) = a^2$$
$$f(-a) = (-a)^2 = a^2 = f(a)$$

EXAMPLE 8. The volume of a sphere is $V = f(r) = \frac{4}{3}\pi r^3$. Its area is $S = g(r) = 4\pi r^2$. Show that $f(r) \div g(r) = \frac{1}{3}r$.

$$\frac{f(r)}{g(r)} = \frac{\frac{4}{3}\pi r^3}{4\pi r^2} = \frac{r}{3}$$

Restating Verbal Relationships as Equations

If a functional relationship is expressed by a verbal statement, it can usually be translated into an equation. The procedure is similar to that used to derive an equation for a verbal problem.

EXAMPLE 9. The cost of a chartered boat is $3.00 per passenger plus an initial fee of $10. Using an equation, express the cost as a function of the number of passengers. Let the independent variable n represent the number of passengers. The charge for n passengers at $3.00 per passenger is $3n$ dollars. Add to this the initial fee of $10, and the total cost is $(3n + 10)$ dollars. If C is the cost, the required relationship is

$$C = 3n + 10$$

Exercise 34

THE FUNCTION CONCEPT

For each of the following pairs of related variables, state which is the dependent variable and which is the independent variable:

1. The thickness of a sheet of paper, t.
   The number of sheets of paper to the inch, n.
2. The speed of a chemical reaction, s.
   The temperature at which the reaction takes place, t.
3. The initial speed of a single-stage rocket fired upward, v.
   The time it requires to reach a given altitude, t.
4. The distance between the earth and the sun, d.
   The time of the year, t.
5. The price of a thoroughbred horse, p.
   The number of races won by the horse, w.
6. The cost of a page of newspaper advertising, c.
   The circulation of the newspaper, n.
7. The variable x.
   The expression $2x^2 + 7x$.
8. The number of bacteria in a culture, n.
   The time during which the bacteria multiply, t.

Choose the one phrase which best completes each of the following statements:

9. The elongation of a coiled spring is a function of (a) its initial length, (b) the amount of weight attached to it, (c) its final length.

10. The number of pencils which can be purchased for one dollar is a function of (a) the size of the pencils, (b) the hardness of the lead, (c) the price per pencil.

11. The number of points scored by a basketball team is a function of (a) the number of players used, (b) the number of games played, (c) the number of field goals and free throws made.

12. The distance covered by a freely falling body is a function of (a) the time it takes to fall, (b) its initial height, (c) its density.

13. The area of a triangle whose base is 8 inches is a function of (a) its perimeter, (b) the size of its vertex angle, (c) its altitude.

14. The number of 1-inch pieces that can be cut from an 8-foot length of copper tubing is a function of (*a*) the thickness of the saw cut, (*b*) the weight of each piece, (*c*) the diameter of the tubing.

15. The time necessary to sail a 15-mile triangular course is a function of (*a*) the velocity of the wind, (*b*) the average speed of the sailboat, (*c*) the direction of the wind.

Find the value of $f(3), f(1), f(0), f(-1), f(-3)$ for each of the following functions:

16. $f(x) = 3x$ **17.** $f(x) = -7x$
18. $f(x) = 4x + 6$ **19.** $f(x) = -5x - 8$
20. $f(x) = x^2$ **21.** $f(x) = -2x^2$
22. $f(x) = 4x^2 + 1$ **23.** $f(x) = -x^2 + x$
24. $f(x) = -3x^2 - 6x + 2$ **25.** $f(x) = x^3$
26. $f(x) = 6 - x^3$ **27.** $f(x) = x - 5x^2$

Show that $f(a) = f(-a)$ for each of the following functions:

28. $f(x) = 3x^2 + 1$ **29.** $f(x) = x^2 - x^4$ **30.** $f(x) = 5 - x^2$

Show that $f(a) = -f(-a)$ for each of the following functions:

31. $f(x) = 8x$ **32.** $f(x) = x^3$ **33.** $f(x) = 2x^3 - 3x$

If $y = \dfrac{2x}{4 - x^2}$, find

34. $f(\frac{1}{2})$ **35.** $f(2a)$ **36.** $f(2a) - f(a)$

37. The circumstance of a circle is given by the formula $C = \pi d$.

   (*a*) What is the independent variable, and what does it represent?

   (*b*) What is the dependent variable, and what does it represent?

   (*c*) If $C = f(d)$, find the value $f(5)$.   What does it represent?

38. The time required to prepare a roast is listed as 20 minutes for each pound of meat and then 15 minutes more.

   (*a*) Write an equation for computing the time *t* required to prepare a roast of *p* pounds.

   (*b*) Evaluate $f(3)$.   What does it represent?

39. A typewriter was sold for $140.   The required down payment was $30.   If the balance was repaid at the rate of $5 per month, write a formula for finding the amount owed at the end of *n* months.

40. The velocity in feet per second of a body falling from rest is 32 times the time of motion expressed in seconds.

   (*a*) Express the relation between time and velocity by means of a formula.

(*b*) Find the value of $f(3)$.   What does it represent?

(*c*) Find the value of $f(4) - f(3)$.   What does this difference represent?

41. The blood pressure of a normal adult, measured in millimeters of mercury, is said to be 110 increased by half the person's age.

(*a*) Write a formula for the blood pressure p of a person of any age n.

(*b*) Evaluate $f(22)$.   What does it represent?

42. An open box is to be formed by cutting squares from the corners of an 18-inch × 24-inch cardboard rectangle and folding along the dotted lines.   The side of the square is x inches.   (See Figure 1.)

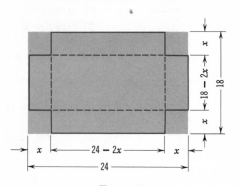

Figure 1

(*a*) What is the domain of the independent variable x?

(*b*) Write a formula for the volume of the box.

(*c*) Evaluate $f(4)$.   What does it represent?

(*d*) Evaluate $f(5)$.   What does it represent?

35   Tabular Representation

A table of values is a familiar method of depicting a relationship between related variables.   It is used extensively to display selected values of functions even in cases where mathematical formulas for the functions are available.   Some common examples of this type of representation are found in tables of compound interest, annuities, powers and roots of numbers, insurance premiums, tube characteristics, logarithms, and trigonometric functions.

The obvious advantages of a table are its simplicity and its convenience of reference. Our interest, however, will be centered on its role as a device for gaining insight into a relationship.

Constructing a Table of Values

A weather observer, concerned with the behavior of the temperature at a given location, assembles data by noting the thermometer readings at regular intervals.   He thus obtains two collections of numbers, namely, the set of time measurements and the set of temperature readings.   To exhibit the relationship between the two sets of numbers, the readings can be arranged in a table as follows:

TABLE 1

Time, x	12 M	2 AM	4 AM	6 AM	8 AM	10 AM	12 N	2 PM	4 PM	6 PM	8 PM	10 PM	12 M
Temperature, y	10	2	−3	−4	6	12	20	28	25	24	22	21	18

It is customary to list the values of the independent variable first in the table, and to list them in ascending order. The time x is the independent variable and is listed in Table 1 since the values of the temperature depend on the time. The corresponding values of the temperature are listed beneath the respective time readings.

Average Rate of Change

Since functional relationships are concerned with changing quantities, it is desirable to introduce a brief and convenient way of expressing changes in variables.

An *increment* of a variable is the difference between two values of the variable, and therefore represents the change in the variable. For the variable x, an increment of x is written as Δx, read "delta x" (in which Δ is not a multiplier of x but is part of a compound symbol). Similarly, Δy is read "delta y" and represents a change in the variable y.

It is obvious from its definition that an increment of a variable is found by the process of subtraction. In fact, we can write

$$\Delta x = x_2 - x_1$$

to indicate that an increment of x is found by subtracting an initial value of x from a second value of x.   It follows that

$$\text{if } x_2 > x_1, \text{ then } \Delta x > 0 \text{ (the increment is positive)}$$
and $\quad$ if $x_2 < x_1$, then $\Delta x < 0$ (the increment is negative)

In Table 1, the value of Δx in each interval is 2.   The value of Δy in the first interval is -8 since the second value, 2, minus the first value, 10, is -8.   In the next interval $\Delta y = -5$ as y changes from 2 to -3, etc.

If the corresponding increments of the dependent variable and independent variable are compared by division, the result is known as the *average rate of change*.   That is,

$$\frac{\Delta y}{\Delta x} = \text{average rate of change of } y \text{ with respect to } x$$

where Δx represents a change in the independent variable and Δy represents the corresponding change in the dependent variable.

The application of this measure is meaningful and important.   For example, considering the first interval of Table 1, we have

$$\Delta x = 2 \text{ hours}$$
$$\Delta y = -8 \text{ degrees}$$

Therefore $\dfrac{\Delta y}{\Delta x} = \dfrac{-8 \text{ degrees}}{2 \text{ hours}} = -4$ degrees per hour

This means that in the interval from 12 midnight to 2 A.M. the temperature was decreasing at an average rate of 4 degrees each hour.   In general the numerical value of $\Delta y/\Delta x$ represents the number of units y changes per unit increase in x.   If $\Delta y/\Delta x$ is positive, then y increases as x increases; if $\Delta y/\Delta x$ is negative, then y decreases as x increases.   The values of Δx, Δy, and $\Delta y/\Delta x$ for each interval of Table 1 are displayed in the following table.

TABLE 2

x	12M	2AM	4AM	6AM	8AM	10AM	12N	2PM	4PM	6PM	8PM	10PM	12M
y	10	2	-3	-4	6	12	20	28	25	24	22	21	18
Δx		2	2	2	2	2	2	2	2	2	2	2	2
Δy		-8	-5	-1	10	6	8	8	-3	-1	-2	-1	-3
$\dfrac{\Delta y}{\Delta x}$		-4	$\dfrac{-5}{2}$	$\dfrac{-1}{2}$	5	3	4	4	$\dfrac{-3}{2}$	$\dfrac{-1}{2}$	-1	$\dfrac{-1}{2}$	$\dfrac{-3}{2}$

Interpolation

It is obviously impossible to list in a table all corresponding values of a relationship which may be of subsequent concern. Therefore a technique for finding or estimating values between those listed in a table is very useful. A process for determining such intermediate values is known as *interpolation.*

Interpolation of tabular values is based on the assumption that the rate of change is constant in the interval under consideration. For example, referring again to Table 1, we note that the temperature at 10 A.M. was 12° and at noon was 20°. What was the temperature reading at 10:30 A.M.?

Lacking other information, it is necessary to assume that the temperature changed at a constant rate between 10 A.M. and 12 noon. In this interval, the average rate of change was

$$\frac{\Delta y}{\Delta x} = \frac{8}{2} = 4 \text{ degrees per hour}$$

At this rate, the change that would occur in $\frac{1}{2}$ hour, that is, between 10 A.M. and 10:30 A.M., is 2°. Since the temperature was 12° at 10 A.M., it is estimated to be 14° at 10:30 A.M.

The work can be arranged as follows:

x	10 A.M.	10:30 A.M.	12 N
y	12°	?	20°

The average rate of change in the interval from 10 A.M. to 12 N is 8/2. The average rate of change in the interval from 10 A.M. to 10:30 A.M. is $\Delta y/\frac{1}{2}$ where Δy is unknown. Since we assume that the rate of change is constant, we have the equality

$$\frac{8}{2} = \frac{\Delta y}{\frac{1}{2}}$$

Solving this equation for Δy, $\Delta y = 2°$. Adding Δy to the given value of y, $12° + 2° = 14°$. Therefore, by interpolation, the temperature at 10:30 A.M. is 14°.

Constructing a Table of Values from an Equation

Although observed values may be used to form a table of values, in many instances such values are obtained by direct computation from an equation of relationship of the variables. The values of the independent variable are arbitrarily chosen (in accordance with the needs of the problem) and the corresponding values of the dependent variable are then computed from the equation.

EXAMPLE 1.   Construct a table of values for the function $y = f(x) = 2x^2 + x$ for intergral values of x in the interval $-3 \leq x \leq +3$.

$f(-3) = 2(-3)^2 + (-3) = 15$.   Therefore when $x = -3$ the corresponding value of y is 15.   Similarly, by evaluating the function for each value of x listed in the table, the corresponding values of y are determined.   The results may be arranged as follows:

x	-3	-2	-1	0	1	2	3
y	15	6	1	0	3	10	21

Exercise 35

TABULAR REPRESENTATION

Construct a table for each of the following functional relationships to indicate the value of y corresponding to $x = -4$, -3, -2, -1, 0, 1, 2, 3, 4, respectively.   In each case, determine the effect on y when x is increased by one.

1. $y = 3x$ **2.** $y = -2x$ **3.** $y = 4x + 2$
4. $y = 4x - 2$ **5.** $y = -3x + 7$ **6.** $y = -3x - 7$

What is the increment of the variable in each of the following instances?

7. A baseball player's batting average changes from 0.321 to 0.304 during the month of July.

8. A car traveling at 46 miles per hour increases its speed to 55 miles per hour.

9. The enrollment of a school was 1348; now it is 2113.

10. A factory expanded its space from 10,500 square feet to 18,000 square feet.

11. The price of a boat was reduced from $880 to $795.

For each of the following relationships, assume that the independent variable is increasing.   Determine whether the dependent variable is then increasing or decreasing and whether the average rate of change is therefore positive or negative.

12. The area of a circle depends on its radius.

13. The volume of a gas depends on the pressure to which it is subjected.

14. The number of apples to a pound depends on the average weight of the apples.

15. The intensity of sound is a function of its distance from the sound source.

16. The electrical resistance of a wire is a function of its diameter.

17. The distance required to stop a car is a function of its speed.

18. The height of a column of mercury depends on its temperature.

19. The time it takes a base runner to reach first base depends on his speed.

20. The length of a steel tape depends on its temperature.

21. In the following table, x represents the length of the edge of a cube in inches and y represents the total area of its 6 faces in square inches. Complete the table.

TABLE 3

x	1	3	5	7	9	11
y	6	54				

(Problems 22–24 refer to Table 3.)

22. What is the average rate of change in the interval from $x = 1$ to $x = 3$?   In this interval, the area increases how many times as fast as the edge?

23. Find $\Delta y/\Delta x$ for the interval from $x = 3$ to $x = 5$.   In this interval, the area increases how many times as fast as the edge?

24. In the interval from $x = 5$ to $x = 7$, for each unit increase of the edge, what is the average number of units increase of the area?

25. In the following table, x represents the length of the edge of a cube in inches, and y represents its volume in cubic inches.   Complete the table.

TABLE 4

x	1	3	5	7	9
y	1	27			

(Problems 26 and 27 refer to Table 4.)

26. Find $\Delta y/\Delta x$ for the first interval of the table.   In this interval, the volume increases how many times as fast as the edge?

27. In the interval from $x = 3$ to $x = 5$, what is the average increase of the volume per unit increase of the edge?

The valve of a water tank is opened at 3 P.M.   The following table shows the number of gallons of water g remaining at each time t.

TABLE 5

t	3:00	3:10	3:20	3:30	3:40	3:50	4:00	4:10
g	450	420	390	360	330	300	270	240

(Problems 28–34 refer to Table 5.)

28. Name the independent and the dependent variable.

29. What is the initial volume of water in the tank?

30. Find Δt for each interval of the table.   Is it constant?

31. Find Δg for each interval of the table.   Is it constant?

32. Find $\Delta g/\Delta t$ for each interval of the table.   Is it constant?

33. Is $\Delta g/\Delta t$ positive or negative?   Interpret the sign.

34. What is the rate of flow of water from the tank in gallons per minute?

For each of the following tables, write an equation which expresses y in terms of x.

35.

x	0	1	2	3	4	5
y	0	4	8	12	16	20

36.

x	1	3	5	7	9	11
y	4	6	8	10	12	14

37.

x	0	1	3	5	6	10
y	10	12	16	20	22	30

38.

x	1	2	5	7	10	12
y	1	4	25	49	100	144

39.

x	-5	-2	0	2	5	8
y	5	2	0	-2	-5	-8

40.

x	1	$1\frac{1}{2}$	2	$2\frac{1}{2}$	3	$3\frac{1}{2}$
y	1	$\frac{2}{3}$	$\frac{1}{2}$	$\frac{2}{5}$	$\frac{1}{3}$	$\frac{2}{7}$

Find the unknown values in each of the following tables by interpolation.

41.

x	3		7
y	9	12	21

42.

w	2	5	10
l		8		32

43.	t	-3	1	8
	v	5		104

44.	u	-7		2
	v	18	0	-36

45.	p	1.0	1.7	2.0
	q	3.46		5.86

46.	n	5		-3
	r	-2.08	-2.56	-4.00

36   Rectangular Coordinates

A graph frequently provides an unusually clear and comprehensive view of the relationship between an independent variable and a function of the variable.   Such representation is most commonly based on the Cartesian system of rectangular coordinates, which was introduced by René Descartes in 1637.   The system involves the following agreements:

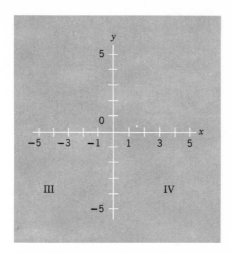

Figure 1

1. A pair of lines are drawn at right angles, one horizontal and one vertical.   These lines are called the *coordinate axes*.   The horizontal line is called the *x-axis* and the vertical line is called the *y-axis*.

2. The plane determined by the coordinate axes is called the *coordinate plane* or the *xy-plane*.

3. The axes divide the coordinate plane into four parts called *quadrants*. They are numbered counterclockwise I, II, III, IV, as in Figure 1.

4. The point of intersection of the axes is called the *origin*.   It is labeled 0 in Figure 1.

5. On each axis, an arbitrary length is selected as a unit for measuring distance. The horizontal distance, measured from the *y*-axis, is called the *x-coordinate*, and the vertical distance, measured from the *x*-axis, is called the *y-coordinate*.

6. Distances measured to the right of the *y*-axis are positive; those to the left are negative. Distances measured upward from the *x*-axis are positive; downward are negative.

7. The location of any point is determined by the value of its *x*-coordinate and *y*-coordinate, or simply, its coordinates. The coordinates of a point *P* are written in the form $P(x, y)$ where *x* represents the *x*-coordinate and *y* represents the *y*-coordinate. The *x*-coordinate is also called the *abscissa* of the point and the *y*-coordinate is called the *ordinate* of the point. The abscissa is always written first in the set of coordinates.

An important consequence of these agreements is that for any one position of a point *P* in the coordinate plane, there exists one and only one pair of coordinates (x, y), and, conversely, for any one pair of coordinates (x, y), there exists a single point *P* in the coordinate plane. Locating a point by means of its coordinates is called *plotting* the point.

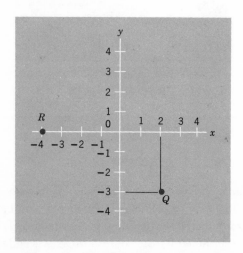

Figure 2

EXAMPLE 1. What are the coordinates of the point *Q* in Figure 2? The abscissa of point *Q* is 2, and its ordinate is −3. Therefore its coordinates are $(2, -3)$.

EXAMPLE 2.   Plot the point $(-4, 0)$ in the xy-plane.   The abscissa is -4, therefore the point is 4 units to the left of the y-axis.   The ordinate is 0, therefore the point is on the x-axis.   The required point is the point R in Figure 2.

EXAMPLE 3.   What are the signs of the coordinates in Quadrant II? All points in the second quadrant are to the left of the y-axis and above the x-axis.   Therefore the abscissa of each point is negative and the ordinate is positive, or $x < 0$, $y > 0$.

Exercise 36

RECTANGULAR COORDINATES

1. Plot the points $A(4, 2)$, $B(-3, -1)$, $C(4, -5)$, $D(-2, 7)$.

2. Plot the points $P(3, 0)$, $Q(3, -3)$, $R(0, 3)$, $S(-3, 3)$, $T(0, -3)$, $U(-3, 0)$, $V(-3, -3)$.

3. Write the coordinates of points E to L, respectively, using the coordinate system illustrated in Figure 3.

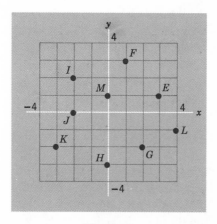

Figure 3

4. Draw a line through the points $(-10, -4)$ and $(14, 8)$.   Which of the following points also lie on this line? $(4, 3)$, $(-4, -1)$, $(0, -2)$, $(0, 1)$, $(8, 5)$, $(-2, 0)$, $(10, 7)$, $(7, 4)$.

5. Plot each set of points. Then join the successive points of each set by straight lines and identify the geometric figures formed.

    (*a*) (0, 0), (8, −2), (9, 2), (1, 4)

    (*b*) (0, 0), (−3, 5), (5, 3)

    (*c*) (0, 0), (14, 4), (5, 7), (−2, 5)

    (*d*) (1, 2), (−3, 3), (0, −2), (4, −3)

    (*e*) (3, 9), (12, 6), (9, −3), (0, 0)

6. Name the quadrant or quadrants for which each of the following statements are true.

    (*a*) x and y have unlike signs.

    (*b*) The abscissa is negative; the ordinate is positive.

    (*c*) $x > 0$

    (*d*) $x > 0, y < 0$

    (*e*) The ordinate is negative.

    (*f*) $y < 0$

    (*g*) The abscissa and ordinate are negative.

7. (*a*) Draw a line through the points (−3, 6) and (5, −10).

    (*b*) List the coordinates of five other points on this line.

    (*c*) Is the ordinate of each of these points −2 times as large as its abscissa?

8. (*a*) Plot the points (3, 5), (3, −2), (3, −7), (3, 0), (3, 8), (3, 3).

    (*b*) Where are all points whose abscissa is 3?

9. (*a*) Plot the points (5, 6), (1, 2), (3, 4), (−5, −4), (0, 1).

    (*b*) Draw a straight line through the plotted points and list the coordinates of several other points on the line.

    (*c*) Express the relationship between the ordinate and the corresponding abscissa of each point in the form of a verbal statement.

    (*d*) Using the relationship developed in part *c*, can you predict whether the point (+183, +184) will lie on the line (extended) through the given points?

    (*e*) Let y represent the ordinate of any of the given points, and let x represent the corresponding abscissa. Write the functional relationship of y and x in the form of an equation.

10. (*a*) Draw a line through the points (8, 4), (2, 1), (0, 0), (−6, −3), (−10, −5).

    (*b*) What are the coordinates of several other points on this line?

    (*c*) Express the relationship of the ordinates and abscissas in the form of a verbal statement.

(d) If the given line were extended, would it pass through the point $(-76, -36)$?

(e) Write an equation which represents the functional relationship between the ordinate (y) and the abscissa (x) of any point on the line.

11. Using the Pythagorean theorem for the relationship of the sides of a right triangle,

(a) Find the length of the line segment joining the points $P(3, 1)$ and $Q(15, 6)$.   See Figure 4.

(b) Prove that the triangle having vertices $(0, 0)$, $(9, -4)$, $(8, 6)$ is scalene.   (A scalene triangle has no two sides equal in length.)

(c) Prove that the triangle whose vertices are $(-4, 3)$, $(0, -3)$ and $(4, 4)$ is isosceles.

(d) Find the length of the line segment joining the points $P_1(x_1, y_1)$ and $P_2(x_2, y_2)$.

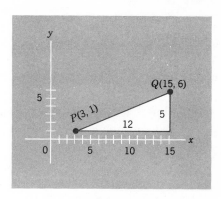

Figure 4

12. Plot each pair of points, locate the midpoint of the line segment joining the pair of points, and determine the coordinates of each midpoint.

(a) $(0, 0)$, $(8, 4)$ (b) $(2, 3)$, $(6, 11)$

(c) $(-1, 0)$, $(7, -6)$ (d) $(-5, 5)$, $(3, 9)$

(e) $(-4, -7)$, $(10, 3)$ (f) $(-1, -2)$, $(-11, 8)$

13. Using the results of Problem 12,

(a) How is the abscissa of the midpoint always related to the abscissas of the given end points?

(b) How is the ordinate of the midpoint always related to the ordinate of the given end points?

(*c*) What is the midpoint of the line segment joining the points (x_1, y_1) and (x_2, y_2)?

14. Plot the points $O(0, 0)$, $A(8, 0)$, $B(0, 6)$ and draw the triangle having these points as vertices.

(*a*) Find the length of the hypotenuse AB.

(*b*) Find the coordinates of the midpoint M of the hypotenuse AB.

(*c*) Find the length of the median OM.

(*d*) How does the length of the median to the hypotenuse of this right triangle compare with the length of the hypotenuse?

(*e*) Repeat these steps for any other right triangle.

(*f*) Complete the statement: The median to the hypotenuse of a right triangle is equal to————the hypotenuse.

37 Graphic Representation

In view of the system of rectangular coordinates for plotting points on a plane, it is possible to represent a functional relationship by means of a graph. The graph of a function defined by $y = f(x)$ is the set of all points (x, y) which satisfy the given equation. The set of points may lie on a straight line. If this is so, we can, of course, draw only a convenient portion of the line. Or the set of points may lie on a curve and then we usually approximate the graph by drawing a smooth curve through the points we choose to plot.

Two types of functions are so important by reason of their frequent application to problems that we shall make special note of their graphic characteristics. They are known as the linear function and the quadratic function.

The Linear Function $y = mx + b$

In the relationship $y = mx + b$, the independent variable is x, the dependent variable is y, and m and b are any constants. Therefore each of the equations $y = 3x + 5$, $y = 2x - 1$, $y = -x$, $y = 2$ is an example of this type of relationship. Let us graph one such example—the function defined by $y = 3x + 5$. Constructing a table of values, we have

x	-2	-1	0	1	2	3	4	5
y	-1		2	5	8	11	14	17	20

These points are plotted on the coordinate system in Figure 1.   They

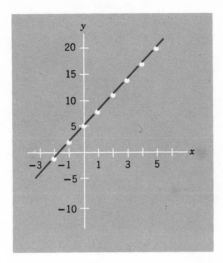

Figure 1

appear to lie on a straight line.   In the study of analytic geometry, it is shown that the graph of every equation $y = mx + b$ is a straight line.

To investigate the role of the constant m in graphing the straight line $y = mx + b$, consider the graphs of the functions defined by

$$y = 3x + 5 \qquad y = 3x - 2$$
$$y = 3x + 12 \qquad y = 3x - 7$$
$$y = 3x$$

The value of m in each equation is 3.   The common property of these lines (Figure 2) is that they are parallel lines.   Therefore the constant m is said to determine the direction of the line $y = mx + b$.

Now consider the functions defined by

$$y = 3x + 5 \qquad y = -3x + 5$$
$$y = x + 5 \qquad y = 5$$

The value of the constant b in each of these equations is 5.   Figure 3 illustrates the graphs.   Each of the lines passes through the common point $(0, 5)$.   This is the point in which each line intersects the y-axis.

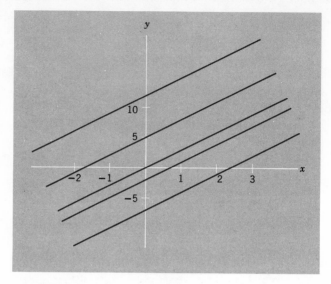

Figure 2

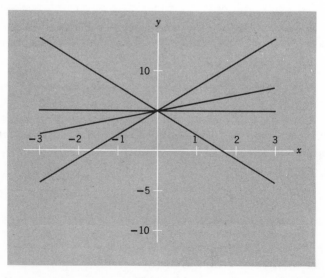

Figure 3

As a result of these considerations, we may list the following important graphic properties of the linear function.

1. The graph of the linear function defined by the equation $y = mx + b$ is a straight line.

2. The constant m determines the direction of the straight line.   It is called the slope of the line.

3. The constant b determines the point at which the line intersects the y-axis.   This point is called the y-intercept.   It is identified by its coordinates $(0, b)$ or simply by its y-coordinate b.

The Linear Function $Ax + By = C$

The equation $4x + 2y = 1$ represents a function of the form $Ax + By = C$.   An equation of this form can be written in the form $y = mx + b$ by solving for y as follows:

$$4x + 2y = 1$$
$$2y = -4x + 1$$
$$y = -2x + \tfrac{1}{2}$$

This is of the form $y = mx + b$.   Therefore the graph of the function $4x + 2y = 1$ is a straight line whose slope is -2 and whose y-intercept is at $(0, \tfrac{1}{2})$.

In the same way, the equation $Ax + By = C(B \neq 0)$ can be written in the form $y = mx + b$.   Again solving for y, we have

$$Ax + By = C$$
$$By = -Ax + C$$
$$y = -\frac{A}{B}x + \frac{C}{B}$$

which is of the form $y = mx + b$.   Therefore the graph of every function of the form $Ax + By = C$ is a straight line.

Since a straight line is determined by any two points, the graph of $Ax + By = C$ can be conveniently drawn by locating two points whose coordinates satisfy the equation.   A pair of points which are frequently used for this purpose are the x- and y-intercepts.   The x-intercept is the point at which the graph intersects the x-axis.   It is found by setting

$y = 0$ and solving the resulting equation for x.   The y intercept can be

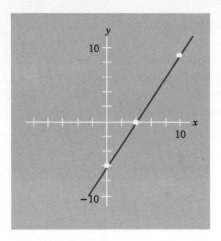

Figure 4

found by setting $x = 0$ and solving the resulting equation for y.   Any third point, if desired, can be plotted to help ensure against errors and to promote accuracy by obtaining points which are spaced sufficiently far apart.

EXAMPLE 1.   Graph the function defined by $3x - 2y = 12$.   When $x = 0$, the resulting equation is $-2y = 12$ from which $y = -6$.

When $y = 0$, the resulting equation is $3x = 12$ from which $x = 4$. These results are shown in the following table.   The third point is plotted as a partial check of the intercepts.   (See Figure 4.)

x	0	4	10
y	-6	0	9

The Quadratic Function $y = ax^2 + bx + c$

We have previously defined $ax^2 + bx + c = 0$ as the quadratic equation and have shown that two and only two values of x will satisfy this equation.   We now remove the restriction that the quadratic expression must equal zero and consider the equality

$$y = ax^2 + bx + c$$

where x is an independent variable whose domain is the set of real numbers. Clearly this introduces an important distinction. Not only does x assume an unlimited number of values, but so also does the dependent variable y. A graph illustrates some important features of this relationship.

EXAMPLE 2.   Plot the graph of the function defined by the equation $y = x^2 - 6x + 5$. The values of x are assigned so as to obtain a desired portion of the curve. The corresponding values of y are then computed from the equation.

x	-1	0	1	2	3	4	5	6
y	12	5	0	-3	-4	-3	0	5

When these points are plotted and joined by a smooth curve drawn through them, we obtain the graph shown in Figure 5. The curve shown in Figure 5 is a *parabola*. It occurs frequently in the arts and sciences. The cables of a suspension bridge form a parabola under certain conditions. The path of an unresisted projectile is usually a parabola, and many reflecting surfaces are designed to have parabolic sections.

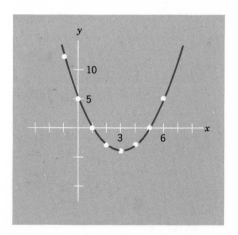

Figure 5

The parabola just drawn can be extended infinitely to the left or right since the domain of x is the set of all real numbers. However, the range of y has a lower limit since the curve has a lowest point at $(3, -4)$. Such a point is called the *vertex* of the parabola. Further, the parabola is symmetrical about the vertical line through the vertex.

EXAMPLE 3.   Graph the function defined by the equation $y = -x^2 + 2x + 3$. The table of values is constructed as before by choosing values of x and computing the corresponding values of y.

x	-2	-1	0	1	2	3	4
y	-5	0	3	4	3	0	-5

This parabola (Figure 6) is open downward, and the vertex (1, 4) therefore is the highest point on the curve).

The important graphic properties of the function $y = ax^2 + bx + c$ illustrated in Examples 2 and 3 are the following:

1. The graph of the function defined by $y = ax^2 + bx + c$ is always a parabola.

2. If $a > 0$, the parabola is open upward, and the vertex is the lowest point on the curve.

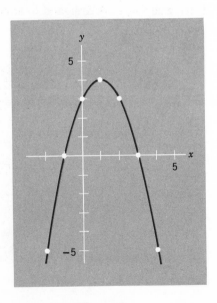

Figure 6

3. If $a < 0$, the parabola is open downward, and the vertex is the highest point on the curve.

4. The parabola is symmetrical about a vertical line drawn through its vertex.

Graphs of Miscellaneous Functions

The methods used for graphing the linear and quadratic functions may be extended to graph other functions. In brief, the following steps are involved:

1. Select values of the independent variable.
2. Using the given equation, compute the corresponding values of the dependent variable.
3. Plot the points thus determined, and draw a smooth curve through the plotted points.
4. To refine the curve between any two points, determine additional points in the interval.

EXAMPLE 4. Graph the function $y = x^3 - 4x$. Choosing zero and positive values of x, we have

x	0	1	2	3	4
y	0	-3	0	15	48

It is evident that as x increases beyond these values, y will also increase.

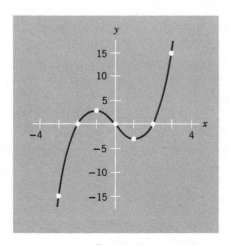

Figure 7

Choosing negative values of x, we have

x	-1	-2	-3
y	3	0	-15

The curve is shown in Figure 7.

Exercise 37

GRAPHIC REPRESENTATION

Graph each of the following equations by plotting three or more points. For each graph, give the slope and the y-intercept by comparing the given equation with the form $y = mx + b$.

1. $y = 2x + 3$ 2. $y = -3x + 1$ 3. $y = \frac{1}{2}x$
4. $y = 5x - 2$ 5. $y = \frac{2}{3}x + 2$ 6. $y = -x - 1$
7. $y = x$ 8. $y = 4x - 4$ 9. $y = 6$

10. From your observations of Problems 1–9,

(a) When the slope is positive, does the value of y increase or decrease as x increases?

(b) When the slope is negative, does y increase or decrease as x increases?

(c) When the slope is zero, what happens to the value of y as x increases?

11. (a) What is the slope of the line $y = 6x - 10$?

(b) Find the coordinates of two points on this line. What is the value of Δx for these two points? Δy? $\Delta y/\Delta x$?

(c) How does the value of $\Delta y/\Delta x$ compare with the slope?

12. (a) What is the slope of the line $y = -3x + 7$?

(b) Find the value of $\Delta y/\Delta x$ for two points on the line.

(c) How does the value of $\Delta y/\Delta x$ compare with the slope?

13. Using the equation $2x + 5y = 20$,

(a) Set $x = 0$, and find the corresponding value of y.

(b) Set $y = 0$, and find the corresponding value of x.

(c) Plot these two points, and draw the graph of the line $2x + 5y = 20$.

(d) When $x = -5$, what is the corresponding value of y? Does the graph you have drawn pass through this point?

Graph each of the following functions by finding the x-intercept, the y-intercept, and one additional point.

14. $x + y = 7$　　　**15.** $2x + y = 10$　　　**16.** $x - y = 6$
17. $2x - y = 4$　　　**18.** $5x - 3y = 30$　　　**19.** $x - 4y = 12$
20. $6x + 5y = 15$　　**21.** $7x - 2y = 20$　　　**22.** $3x - 2y = -18$

Plot each pair of equations on the same set of axes.　How are the lines of each pair related?

23. $2x + y = 6$　　　**24.** $3x - 2y = 12$　　**25.**　$x - 5y = 5$
　　　$4x + 2y = 20$　　　　　$9x - 6y = 24$　　　　　$4x - 20y = -20$
26.　$x + y = -3$　　**27.**　$7x - 5y = 70$　　**28.**　$6x - y = 12$
　　　$5x + 5y = 30$　　　　$14x - 10y = 70$　　　　$18x - 3y = -18$

Plot the following pairs of equations.　How is each pair of lines related?

29.　$x + y = 6$　　**30.**　$5x - 3y = 15$　　**31.** $2x - y = -14$
　　　$2x + 2y = 12$　　　　$25x - 15y = 75$　　　$6x - 3y = -42$

32. Given the quadratic function $y = x^2 - 8x + 12$.

(a) Complete the table of corresponding values:

x	0	1	2	3	4	5	6	7	8
y									

(b) Plot the points listed in the table and draw the parabola.

(c) Find the coordinates of the x-intercepts, y-intercept, and vertex.

(d) Solve the quadratic equation $x^2 - 8x + 12 = 0$.
How are the roots of this equation related to the x-intercepts of the given function?

33. Given the function $y = -x^2 + 8x - 15$.

(a) Complete the table by computing the corresponding values of y.

x	1	2	3	4	5	6	7
y							

(b) Plot the points listed in the table and draw the graph.

(c) Solve the quadratic equation $-x^2 + 8x - 15 = 0$.

(d) How are the roots of the equation $-x^2 + 8x - 15 = 0$ related to the x-intercepts of the function $y = -x^2 + 8x - 15$?

For each of the following, complete the given table and graph the function.　In each example, list the coordinates of the vertex.

34. $y^2 = x^2 - 4x$

x	-1	0	1	2	3	4	5
y							

35. $y = x^2 - 16x + 60$

x	2	4	6	8	10	12
y						

36. $y = x^2 - 4$

x	-3	-2	-1	0	1	2	3	4
y								

37. $y = x^2 - 6x - 7$

x	-1	0	1	2	3	4	5	6	7	8
y										

38. $y = x^2 + 2x - 3$

x	-5	-4	-3	-2	-1	0	1	2	3
y									

39. $y = -x^2 + 2x + 8$

x	-3	-2	-1	0	1	2	3	4	5
y									

Graph enough of each of the following functions to locate the vertex. List the coordinates of the vertex for each parabola.

40. $y = x^2 - 2x + 1$ **41.** $y = x^2 - 8x + 4$

42. $y = x^2 + 4x - 6$ **43.** $y = 2x^2 - 8x$

44. $y = -3x^2 + 6x + 1$ **45.** $y = x^2$

46. $y = -x^2 + 10x + 5$ **47.** $y = -2x^2 - 12x + 9$

Graph the following functions:

48. $y = \dfrac{12}{x}$ **49.** $y = -\dfrac{12}{x}$

50. $y = x^3$ **51.** $y = -x^3$

52. $y = x^3 - x^2$ **53.** $y = x^3 + 3x^2 + 2x$

54. $y = x^4$ **55.** $y = -x^4$

38 Graphic Solution of Systems of Equations

We have just seen that a graph contains all points whose coordinates satisfy a given equation.   Conversely, each solution of a given equation corresponds to a point which is on the graph of the equation.   This consideration is the basis for a graphic solution of a given system of equations.

If two linear equations in two unknowns are plotted on the same axes, the coordinates of the point of intersection will satisfy both equations and therefore represent their common solution. Although the method is sometimes less direct than an algebraic solution (Section 18), and frequently produces no more than an approximation of the solution, it nevertheless has two noteworthy virtues:

1. The graphic method quickly and clearly establishes the number of real solutions of a given system of equations; and

2. The graphic method can be readily extended to systems of higher degree equations whose solution by algebraic methods may be difficult or highly involved.

Systems of Two Linear Equations

Let us solve graphically the system

$$(1) \quad 3x + 2y = 12$$
$$(2) \quad x - y = 9$$

The x-intercept of the line defined by Equation 1 is 4; the y-intercept is 6. For the line defined by Equation 2, the x-intercept is 9, and the y-intercept is -9. The graphs of these two equations are shown in Figure 1. The point $(6, -3)$ is the point of intersection of the two lines. It lies on both lines, and its coordinates satisfy both equations. Therefore the common solution of the system is $x = 6$, $y = -3$.

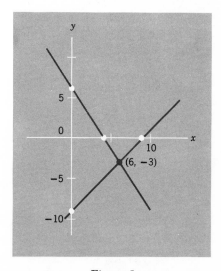

Figure 1

Now consider the general problem of the number of common solutions of the system

$$a_1x + b_1y = c_1$$
$$a_2x + b_2y = c_2$$

The graphic method tells us that these equations will represent two straight lines which may be parallel, may intersect, or may coincide. These are the only possibilities. Therefore this system will have (*a*) one and only one solution if the lines intersect (because two lines can intersect at one and only one point), or (*b*) no solutions if the lines are parallel (because parallel lines have no points in common), or (*c*) an infinite number of solutions if the lines coincide (since all points on coincident lines are common to both lines).

EXAMPLE 1.   Investigate the common solutions of the system

$$2x + y = 14$$
$$4x + 2y = 28$$

The *x*-intercept of each equation is 7, and the *y*-intercept of each equation is 14. Since the two lines have the same intercepts, they coincide. There are an infinite number of common solutions, any number of which can be determined by assigning a value to *x* and solving for the corresponding value of *y*.   (See Figure 2.)

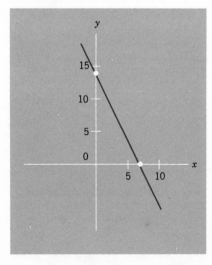

Figure 2

EXAMPLE 2.   Investigate the common solutions of the system

$$(1) \quad x - 2y = 10$$
$$(2) \quad 3x - 6y = 12$$

For Equation 1

x	0	10
y	−5	0

For Equation 2

x	0	4
y	−2	0

The graphs of these equations are shown in Figure 3.   Since the lines are parallel, the given system has no common solution.   (One way to

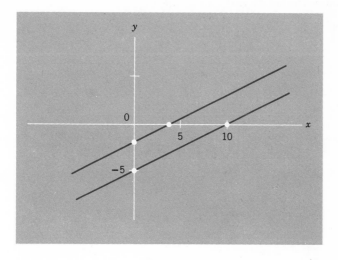

Figure 3

verify that the lines are parallel is to show that the slopes are equal by writing each equation in the form $y = mx + b$.   What are some other ways?)

Systems of One Quadratic and One Linear Equation

The graph of the quadratic function $y = ax^2 + bx + c$ is a parabola, and the graph of the linear function $y = mx + b$ or $Ax + By = C$ is a straight line.   In general, the number of points which a parabola and a line may have in common is two, one, or none.   The following diagrams illustrate these possibilities.

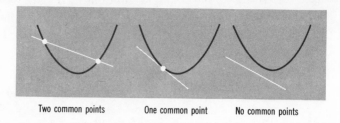

Two common points   One common point   No common points

Figure 4

It follows, therefore, that the system may have two, one, or no common solutions in the set of real numbers.

EXAMPLE 3.   Solve the system of equations.

$$(1) \quad y = x^2 - 6x + 9$$
$$(2) \quad x - y = -3$$

To plot the given pair of equations, we construct a table of values for each:

For Equation 1									For Equation 2		
x	0	1	2	3	4	5	6	7	x	0	-3
y	9	4	1	0	1	4	9	16	y	3	0

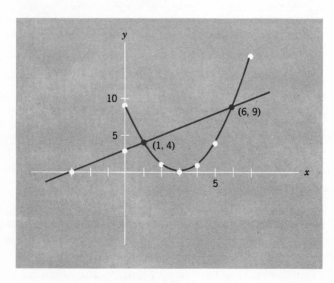

Figure 5

The points of intersection are at $(1, 4)$ and $(6, 9)$. Therefore there are two common solutions, namely, $x = 1$, $y = 4$, and $x = 6$, $y = 9$.

Exercise 38

GRAPHIC SOLUTION OF SYSTEMS OF EQUATIONS

Plot the graphs of the following systems of equations. In each example, determine whether the system has zero, one, or an infinite number of solutions. Where the system has a single common solution, find its value graphically.

1. $x + y = 5$
   $2x - 4y = 4$
2. $x + y = 8$
   $x - y = 6$
3. $x - y = 9$
   $x + y = 7$
4. $2x + y = 6$
   $6x - 3y = 18$
5. $4x - 3y = 6$
   $8x - 6y = 12$
6. $x + 2y = 11$
   $2x + y = 10$
7. $5x + 2y = 10$
   $15x + 6y = 60$
8. $3x - y = 9$
   $4x - y = 15$
9. $x + 7y = 21$
   $-x + y = 11$
10. $x + y = 13$
    $x - y = 13$
11. $x - y = 7$
    $2x - 2y = 7$
12. $x + 4y = 1$
    $x + 5y = 0$
13. $y = x + 3$
    $y = 11 - x$
14. $y = 2x - 1$
    $3x + y = 2$
15. $y = -2x$
    $2x + y = 6$
16. $y = 2x - 3$
    $4x = 2y + 6$
17. $2x - 3y = 15$
    $x + y = 0$
18. $3x + y = -2$
    $x + 3y = 26$

Solve the following systems graphically. Where necessary, estimate the solution to the nearest tenth.

19. $y = x^2 - x - 6$
    $y = -2x$
20. $y = x^2 - 9x + 14$
    $y = x - 2$
21. $y = -x^2 + 4x$
    $2x + y = 9$
22. $y = x^2 - 5x - 6$
    $y = 0$
23. $y = x^2 - 2x - 5$
    $y = 0$
24. $y = x^2 + x - 2$
    $y = x + 1$
25. $y = -x^2 + 7x - 5$
    $y = 2x + 2$
26. $y = x^2 - 6x$
    $x - y + 8 = 0$
27. $y = x^2 - x - 12$
    $x - y = 1$

28. Find the real roots of the equation $x^3 - 3x^2 + 4 = 0$ by solving graphically the system $y = x^3 - 3x^2 + 4$; $y = 0$.

29. Find the real roots of the equation $x^3 + x^2 - 6x = 0$ by solving graphically the system $y = x^3 + x^2 - 6x$; $y = 0$.

30. Solve graphically the system $y = x^3$; $8x - 7y + 40 = 0$.

31. Given the system of equations $y = x^2 - 2x + 10$; $y = 0$.

(*a*) Show, by plotting the graphs, that the system has no solution in the set of real numbers.

(*b*) Show, by solving the quadratic equation $x^2 - 2x + 10 = 0$, that the system has two solutions in the set of complex numbers.

39 Graphic Solution of Problems in Maxima and Minima

Among the many simple applications of the analysis of a function, one of the most important is the determination of its maximum or minimum values.   These terms are used to designate the largest or smallest element among a carefully specified set of elements.   The problem has broad practical value for engineers, scientists, economists, manufacturers, and others who may seek the one element of a given set which represents, for example, a least cost, or greatest area, or highest altitude, or shortest distance.

We have previously noted that the vertex of a parabola is its highest point or lowest point accordingly as the parabola opens downward or upward.   This means that a function of the form $y = ax^2 + bx + c$ has a maximum value if $a < 0$ and a minimum value if $a > 0$.   Now let us apply this information to a problem situation.

EXAMPLE 1.   A farmer has 600 feet of fencing with which he wishes to enclose a rectangular field along a straight river bank.   If no fence is needed along the river bank, what dimensions will provide a rectangle of greatest area?   Let x be the width of the rectangle in feet, then $600 - 2x$ is its length in feet (Figure 1).   Also, let A be the area of the rectangle in

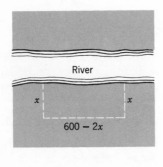

Figure 1

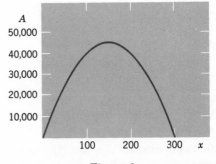

Figure 2

square feet.   The area of a rectangle is the product of its length and width, therefore

$$A = x(600 - 2x) = -2x^2 + 600x$$

This equation is of the form $y = ax^2 + bx + c$.   Since $a < 0$, the graph is a parabola opening downward, and its vertex is a maximum point (Figure 2).   The coordinates of the vertex are $(150, 45{,}000)$.   Thus the greatest area which can be enclosed by 600 feet of fencing under the stated conditions is 45,000 square feet.   The width x necessary to obtain this area is 150 feet, the corresponding length is 300 feet.

Some graphs may have a point which is the highest or lowest point in its immediate vicinity, even though it is not the highest or lowest point on the entire curve.   For example, in Figure 3, point A is the highest

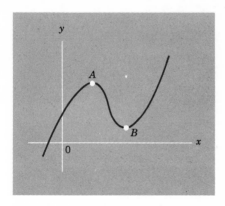

Figure 3

point in its immediate vicinity, and point B is the lowest point in its vicinity.   Obviously, such points will determine the maximum or minimum value of a function in a given interval, and can be located by plotting the graph in that interval.   In practice, the conditions of a problem will prescribe the domain of the independent variable in which the maximum or minimum occurs.

EXAMPLE 2.   A manufacturer is asked to produce an open-top box from a 6 inch square piece of metal by cutting out equal squares from the corners and folding up the sides (Figure 4).   What length of side of the cut-out squares will make the volume of the resulting box a maximum? Let x be the length, in inches, of the side of the cut-out squares.   Then x will be the height of the resulting box, and $6 - 2x$ will be the length of the edge of the square base.   The volume of the box is the product of the area of the base and the height and is therefore given by

$$V = x(6 - 2x)^2 = 4x^3 - 24x^2 + 36x$$

Thus the volume is a function of the length x.   The graph of this function is shown in Figure 5.   The domain of x which has meaning in this

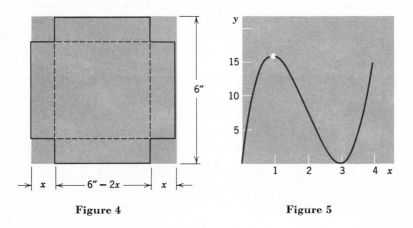

Figure 4   Figure 5

problem is $x = 0$ to $x = 3$.   (Why?)   In this interval, the function has a maximum at $x = 1$.   Therefore, if a 1 inch square is cut out from each corner, the box will have its greatest volume, namely, 16 cubic inches.

Exercise 39

GRAPHIC SOLUTION OF PROBLEMS IN MAXIMA AND MINIMA

1. Graph the function $y = x^2 - 4x + 15$, and locate its vertex.

(a) For what value of x will the expression $x^2 - 4x + 15$ have the least value?

(b) What is the least value the quantity $x^2 - 4x + 15$ may have?

2. Graph the function $y = -x^2 + 6x + 1$, and locate its vertex.

(a) For what value of x will the expression $-x^2 + 6x + 1$ have the greatest value?

(b) What is the greatest numerical value of the quantity $-x^2 + 6x + 1$?

For each of the following functions, plot the graph, and from the graph, estimate the maximum or minimum value of the function.   State whether each value is a maximum or minimum.

3. $y = x^2 - 6x + 1$ **4.** $y = x^2 + 2x - 5$
5. $y = -x^2 + 8x + 9$ **6.** $y = -3x^2 - 6x + 2$
7. $y = 2x^2 - 10x$ **8.** $y = 7 - 5x^2$
9. $y = x^2 - 3x - 2$ **10.** $y = 10 + 3x - 3x^2$

Determine the value or values of x for which each of the following functions has a maximum or minimum point:

11. $y = x^3 - 3x^2$ **12.** $y = 12x - x^3$
13. $y = x^3 - x^2 - x$ **14.** $y = x^4 - 2x^2 + 1$

15. Find the value of x for which the product $x(x + 2)$ will be a minimum.

16. Find the value of x for which the product $x(6 - x)$ will be a maximum.

Without graphing, determine which of the following products has a maximum value and which has a minimum value.   Explain your answer.

17. $(x + 3)(x - 5)$ **18.** $(5 - x)(4 + x)$
19. $(x - 1)(3 + x)$ **20.** $(x + 4)(7 - x)$
21. $(2 - x)(3 - x)$ **22.** $(x - 1)(x - 2)$

23. Separate 12 into two numbers whose product is a maximum.

(a) Let x be one number, and express the other number in terms of x.

(b) Write the product y as a function of x.

(c) Graph the function, and find the value of x for which the function has a maximum.

24. Find the area of the largest rectangle that has a perimeter of 30 feet.

(a) Let x be the width of the rectangle, and express the length in terms of x.

(b) Write the area A as a function of x.

(c) Graph the function, and determine its maximum value.

25. A rectangular gutter is formed by folding up the sides of a strip of tin 12 inches wide. To carry the greatest amount of water, the cross section area of the gutter must be a maximum. How wide and how deep must the gutter be made to have a maximum carrying capacity?

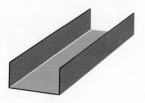

Figure 6

26. A rectangular box is made from a sheet of metal 4 feet by 8 feet by cutting out equal square pieces from the corners and turning up the projecting portions which remain. Find, approximately, the edge of the square which will lead to the largest possible volume.

27. The height of a ball thrown upward with an initial speed of 48 feet per second is given by the formula $h = 48t - 16t^2$, where the independent variable t is measured in seconds and the function h is measured in feet. What is the greatest height to which the ball rises?

28. An open-top box is to have a square base and a volume of 6 cubic feet. The cost of material for the base is 10 cents per square foot and for the sides is 6 cents per square foot. Find the approximate dimensions to make the cost of materials a minimum.

29. The average circulation of the Sunday edition of a town newspaper is 1100 copies when it sells for 15 cents per copy. It is estimated that the circulation will increase by 100 for each penny decrease in price. What price will provide the maximum gross receipts for the newspaper?

40 Inequalities in Two Variables

In Figure 1, the graph of the linear function $y = \frac{1}{2}x$ is shown. At each point A, B, C, and D, of the graph, the ordinate y equals half of the abscissa x. What may be said of the points A', B', C', and D'? The abscissa of each of these points is the same as the abscissa of the corresponding points A, B, C, and D, but the ordinate of each of these points

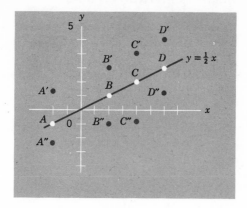

Figure 1

is greater than the ordinate of A, B, C, and D.   Since this is true for all points above the line $y = \frac{1}{2}x$, for all such points

$$y > \frac{1}{2}x$$

In the same manner, for all points below the line, such as A'', B'', C'', and D'',

$$y < \frac{1}{2}x$$

The line $y = \frac{1}{2}x$ therefore divides the xy-plane into two parts, shown in Figure 2.   For the part of the plane above the line, $y > \frac{1}{2}x$, and for the part of the plane below the line, $y < \frac{1}{2}x$.   It is evident, for example, that for the point $E(4,3)$ $y > \frac{1}{2}x$, since $3 > 2$, and that for the point $F(4,1)$ $y < \frac{1}{2}x$, since $1 < 2$.

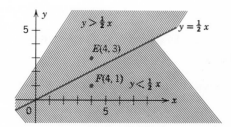

Figure 2

Solution of Inequalities in Two Variables

Expressions such as $y > \frac{1}{2}x$ and $2y + 3x > 12$ are called inequalities in two variables, and the postulates of inequalities (Section 17) apply to their solution.   We illustrate with several examples.

EXAMPLE 1.   Solve the inequality $2y - x > 4$.

$$2y - x > 4 \qquad \text{The given inequality}$$
$$2y > x + 4 \qquad \text{Additive property}$$
$$y > \frac{x}{2} + 2 \qquad \text{Multiplicative property}$$

Therefore whenever $y > x/2 + 2$, the given inequality is satisfied.   As a check, we may choose the values $x = 2$, $y = 4$, where $y > x/2 + 2$. Substituting these values in the original inequality, $2y - x > 4$, we have $8 - 2 > 4$, showing that $x = 2$, $y = 4$ is one solution of the inequality. We note also that in the graph of this inequality, Figure 3, the point (2,4)

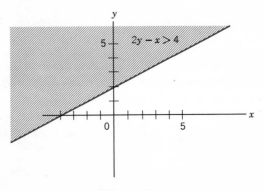

Figure 3

is above the line.   Note that all points in the shaded area above the line satisfy the inequality.

EXAMPLE 2.   Solve the inequality $x + y - 2 < 0$.

$$x + y - 2 < 0 \qquad \text{The given inequality}$$
$$y < 2 - x \qquad \text{Additive property}$$

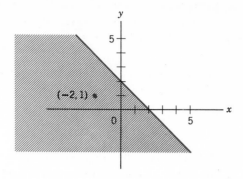

Figure 4

Therefore whenever $y < 2 - x$, the inequality is satisfied. As a check, we may choose the values $x = -2$, $y = 1$, where $y < 2 - x$. Substituting these values of x and y in the given inequality, we have $-2 + 1 - 2 < 0$, or $-3 < 0$, showing that $x = -2$, $y = 1$, is one solution of the inequality. In Figure 4 we note that the point $(-2, 1)$ is in the shaded area below the line, and in fact all points in this area satisfy the given inequality.

The Graph of an Inequality in One Variable

The solution of an inequality in one variable may be strikingly pictured by a graph. Let us consider again the problem of the salesman's choice

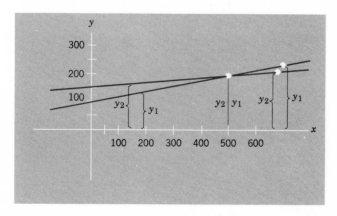

Figure 5

between two jobs, discussed in Section 17.   In this problem a salesman had a choice of two jobs, one in which he was offered a fixed salary of $100 plus a commission of 20% of all sales, and the other in which the fixed salary was $150 plus a commission of 10% of all sales.   If we use x and y instead of s and i to represent the sales and income (it really does not matter what letters we use) then

$$y_1 = 100 + 0.20x$$
$$y_2 = 150 + 0.10x$$

The two equations are graphed on the same axes, as in Figure 5.   Here we see that when x, the amount of sales, is greater than 500, y_1, the income

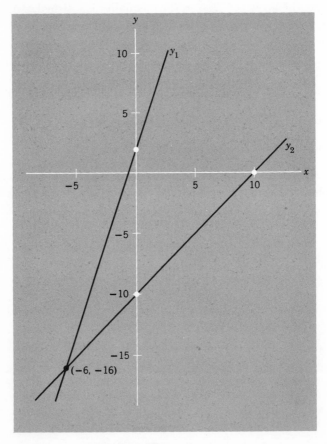

Figure 6

from the first plan, is greater than y_2, the income from the second plan. We see also that when the sales are exactly \$500, the income is the same for both plans, and, finally, when the sales are less than \$500, the second plan provides the higher income.   As a final example, let us solve and graph the inequality

$$3x + 2 > x - 10$$

Adding $-x - 2$ to both members, we get

$$2x > -12$$
from which $$x > -6$$

Therefore for all values of x greater than -6, the inequality is satisfied.
   To graph the inequality, we let

$$y_1 = 3x + 2$$
$$y_2 = x - 10$$

and draw the graph of each equation, as in Figure 6.   We see that when $x > -6$, y_1 is above y_2, so that $y_1 > y_2$, or $3x + 2 > x - 10$.

Exercise 40

INEQUALITIES IN TWO VARIABLES

Solve each of the following inequalities for y in terms of x.   Draw the graph and indicate by shading the part of the xy-plane for which the inequality is true.

1. $3x - 2y + 4 > 0$ **2.** $y + 2 < x$ **3.** $2x - 3y > 6$
4. $y - x > 3$ **5.** $x + y > 4$ **6.** $y + 7 > 2x$
7. $y - 1 < 2x$ **8.** $x - y - 4 < 0$ **9.** $y + 3x - 6 > 0$
10. $6x + 4y > 12$ **11.** $y + x < 2x$ **12.** $y - x < 3x$

Solve each of the following inequalities in one variable, and check the solution by drawing the graph of each member, as in Figure 5.

13. $x + 7 > 2x - 1$ **14.** $3x - 5 > -6x + 4$ **15.** $3x - 2 > 2x + 1$

16. $2x + 1 < x - 3$ **17.** $\dfrac{2x - 3}{2} > 5$ **18.** $\dfrac{3x - 1}{3} < 2x + 1$

19. Describe the part of the xy-plane where the inequality $x > y - 2$ is satisfied.   Note that x is solved in terms of y.
20. Describe the part of the xy-plane where the inequality $x < y - 2$ is satisfied.   Note that x is solved in terms of y.

CHAPTER 8

Measurement

Exact and Approximate Numbers

If we should count 25 students in a certain room, and our count is correct, the number 25 is an exact number. If we should measure the height of a particular student, however, and find it to be 68.5 inches, the number 68.5 is not exact in the same sense that the number 25 is exact. Slight errors in our measurement due to our personal limitations as well as to imperfections in our measuring instrument prevent our measurement from being absolutely accurate, that is, exact. We say that 68.5, obtained by measurement, is an *approximate number*. In fact, all numbers obtained by measurement are approximate. In contrast, numbers obtained by counting are exact.

Direct and Indirect Measurement

If we wish to find the number of square yards in a rectangular carpet, we first measure the length and the width with some kind of measuring instrument, such as a yardstick. This is a *direct measurement;* we are comparing the number of yards in the length and width of the carpet with a standard unit—the yard. To find the area of the carpet, we use the formula $A = lw$; this is an *indirect measurement*. Since the length and width of the carpet are approximate numbers, the area measurement will also be an approximate number. Another example of an indirect measurement is that obtained by an ordinary spring balance. When we weigh a fish on a spring scale, we are not comparing its weight with a standard weight, but are making use of Hooke's Law, which states that the stretch of the spring is proportional to the weight applied.   On the other hand, when a chemist weighs a sample on a balance, he generally will compare the weight of the sample with standardized weights, and therefore makes a direct measurement.   Thus we see that the weight of an object can be determined either directly or indirectly.   Similarly, the area of a rectangle can be determined either indirectly by the use of the formula $A = lw$, or directly by noting the number of times a unit square is contained in the given area.   It is evident that the direct measurement of an area, although theoretically possible, is inconvenient and impractical.

Exercise 41

APPROXIMATION IN MEASUREMENT

State which of the following are likely to be exact numbers and which are likely to be approximate numbers:

1. 3 dozen pencils
2. 3 inches
3. 4 hours 27 minutes
4. 5.3 miles
5. $2.75
6. $\frac{1}{2}$ dozen eggs
7. $1\frac{1}{2}$ pounds of butter
8. 65 miles per hour
9. 98.6°F.
10. 832 students
11. 15,000 football fans
12. 16 calories per slice of toast
13. The area of Illinois is 56,400 square miles
14. 2.54 centimeters equal 1 inch
15. 10 cans of tomato juice
16. 10 cups of flour
17. 1 lump of sugar
18. 2 teaspoons salt
19. 4 quart bottles of milk
20. 15 gallons of gasoline
21. π
22. 3.1416
23. 125 kilowatt hours
24. 3 cubic feet of gas per minute
25. Humidity of 42%

State which of the following are indirect measurements and which are direct measurements:

26. Drawing a 60° angle by a protractor.
27. Drawing a 60° angle by constructing an equilateral triangle.
28. Reading a temperature from an ordinary room thermometer.

29. Measuring the length of a lot.
30. Measuring the distance to the moon.
31. Finding the height of the ceiling in a room.
32. Weighing a pound of sugar on a grocer's scale.
33. Finding the area of a triangle.
34. Finding the perimeter of a square.
35. Finding the perimeter of a scalene triangle.
36. Finding the perimeter of an equilateral triangle.

42   Accuracy and Precision of Approximate Numbers

Errors in Approximate Numbers

Let us again consider the height of the student previously stated to be 68.5 inches.   If 68.5 is an approximate number, then obviously there is some margin of error to be considered.   If, for example, we had some way of knowing that the true value of the measurement was 68.6 inches, our measurement would be in error by 0.1 inch.   But clearly we can never know what the true value is since all measurements are approximate. When we state that the height of the student is 68.5 inches, we are indicating that the true height lies between 68.45 inches and 68.55 inches; our measurement of 68.5 inches is correct to the nearest tenth of an inch. The maximum error is 0.05 inch.   In general, the maximum error of a measured number is one-half the smallest place value in which it is expressed.

EXAMPLE 1.   The maximum error in the number 167 miles is 0.5 mile. The true value, therefore, lies between 166.5 miles and 167.5 miles.

EXAMPLE 2.   The maximum error in the number 3.014 inches is 0.0005 inch.   The true value therefore lies between 3.0135 inches and 3.0145 inches.

Absolute and Relative Errors

If a student misses two problems in an examination, we have no measure of his success unless we know the total number of problems in the examination.   It is one thing if he misses two problems out of one hundred; quite another if he misses two out of five.   In either example, if we consider each problem missed as an error, the *absolute error* is 2 in each case.   The *relative error* in the first example, however, is 2/100; in the second example, it is 2/5.   Expressed as percentages, these figures are 2 per cent and 40 per cent, and we have some indication of the success of the student.   We may define the absolute error of a measurement as the difference between the approximate and the true value of the measurement.   Since, however, we never know the true value, we consider the absolute error to be the maximum error as defined in the previous paragraph.   We define the relative error as the absolute error divided by the measurement.   The following examples should make clear the meaning of the two terms.

Measurement	Absolute Error	Relative Error
4.27 miles	0.005 mile	$\dfrac{0.005}{4.27} = 0.0012$
3.217 grams	0.0005 gram	$\dfrac{0.0005}{3.217} = 0.00012$
9 feet	0.5 foot	$\dfrac{0.5}{9} = 0.06$

Significant Digits

An important concept in measurement is that of *significant digits*. The significant digits of a measured number are defined according to the following rules.

1. All nonzero digits are significant.   The digits in 12.34 are all significant; the number 12.34 has four significant digits.

2. All zeros which lie between nonzero digits are significant.   The digits in 120.5 are all significant; the number 120.5 has four significant digits.

3. All zeros following a nonzero digit and to the right of the decimal point are significant.   The digits in 12.700 are all significant; the number 12.700 has five significant digits.

4. Zeros not preceded by a nonzero digit are not significant, since they are used only to locate the decimal point.   In the number 0.0043, only the digits 4 and 3 are significant; the number 0.0043 has two significant digits.

5. Consecutive zeros which terminate an integer are usually not significant.   In the number 240,000 miles, the approximate distance to

the moon, only the digits 2 and 4 are significant; the number 240,000 has only 2 significant digits.   On the other hand, if exactly 2000 cars were produced in one day, then 2000 has four significant digits.

In summary, if each digit of an approximate number except the last one is exact, and if the error in the last digit is not more than one-half the unit of measurement, then each digit of the number is significant. Thus nonzero digits are always significant, and a zero in an approximate number is significant if and only if it is needed to express the count of the number of times the given number contains the unit of measurement. The following listing of numbers and their significant digits illustrate the usual interpretation of this principle:

> 2.001 inches, 2, 0, 0, 1
> 0.020 inch, 2, and the final 0
> 500 cubic centimeters, 5
> 500.0 cubic centimeters, 5, 0, 0, 0
> 37.45 miles, 3, 7, 4, 5

Accuracy and Precision of Approximate Numbers

Although accuracy and precision are considered as synonymous words in everyday language, they have quite distinct meanings in mathematics. The *precision* of a measured number depends on the *absolute error;* two numbers are equally precise if they have the same absolute error, and one number is more precise than another if its absolute error is smaller. The *accuracy* of a measured number depends on the *relative error;* two numbers have equal accuracy if they have the same relative error, and one number is more accurate than another if its relative error is smaller.

The precision of measurements expressed in the same unit depends on the number of decimal places.   The measurements 28.45 inches and 0.27 inch are equally precise, that is, they are precise to the nearest hundredth of an inch, and their absolute errors are equal.   On the other hand, the accuracy of measurements depends only on the number of significant digits, since the relative error is independent of the unit of measurement.   The measurement 23.45 miles is more accurate than the measurement 0.438 inch, since it contains more significant digits and therefore has a smaller relative error.

EXAMPLE 1.   27.32 inches and 0.16 inch have the same degree of precision, but 27.32 inches is more accurate than 0.16 inch.

EXAMPLE 2.   4.58 ounces is more precise than 35.2 ounces, but they both have the same degree of accuracy.

EXAMPLE 3.   32.1 miles and 20.3 miles have the same degree of accuracy and precision.

Exercise 42

ACCURACY AND PRECISION OF APPROXIMATE NUMBERS

State the absolute and relative errors of each of the following numbers. State also between what two numbers each measurement lies.

1. 2.54 pounds	**2.** 28.45 miles	**3.** 8.7 inches
4. 98.6°F.	**5.** 186,200 miles per second	**6.** 2.5 hours
7. 5280 feet	**8.** 258 horsepower	**9.** 12.4 bushels
10. 842 gallons	**11.** 0.3937 inch	**12.** 28.875 cubic inches
13. 3.1416 inches	**14.** 0.0245 gram	**15.** 0.00025 inch

List the significant digits in each of the following numbers:

16. 93,000,000 miles	**17.** 240,000 miles	**18.** 0.0030 inch
19. 0.003 inch	**20.** 1.003 inches	**21.** 18.43 cubic feet
22. 0.107 centimeter	**23.** 2482 cars	**24.** 2000 spectators
25. $1.04	**26.** 0.08 centimeter	**27.** 2000 cars delivered
28. 7400 pounds	**29.** 6.000 ounces	**30.** 6 ounces

Which of the following numbers have the same degree of precision as 7.032 inches?

31. 3.40 inches	**32.** 3.45 inches	**33.** 0.006 inch
34. 1.003 inches	**35.** 254 inches	**36.** 2540 inches
37. 2.504 inches	**38.** 73.000 inches	**39.** 876.001 inches

Which of the following numbers have the same degree of accuracy as 3.0012 centimeters?

40. 30012 centimeters	**41.** 312 meters	**42.** 0.0020 meter
43. 1.0024 inches	**44.** 0.014215 inch	**45.** 628.45 yards
46. 3.2100 centimeters	**47.** 32,100 feet	**48.** 321 centimeters
49. 60102 meters		

43   Computation with Approximate Numbers

We have discussed the errors inherent in numbers obtained by direct measurement. When approximate numbers are used in calculations, the results will also be subject to errors. Suppose we find the dimensions of a rectangular field to be 35.67 feet by 86.72 feet. Then the area is 35.67 × 86.72 or 3093.3024 square feet. Clearly, if we give the latter figure as the result, we are "either exhibiting our ignorance or our desire to create a false impression."* For the number 3093.3024 has eight significant digits, which indicates a much greater degree of accuracy than that of the measurements from which it was obtained—a most unlikely result. We shall therefore discuss certain rules for obtaining calculated results consistent with the degree of accuracy or precision of the direct measurements used in making the calculations.

Rounding Off Numbers

The process of rounding off numbers is necessary in calculations with approximate numbers. The following examples will show how numbers are rounded off.

EXAMPLE 1.   The number 23.473, rounded off to four significant digits, is 23.47. Since the last digit, 3, is less than 5, we simply drop it; 23.473 is closer to 23.47 than to 23.48.

EXAMPLE 2.   The number 4.728, rounded off to three significant digits, is 4.73. Since the last digit, 8, is greater than 5, we increase the preceding figure by 1; 4.728 is closer to 4.73 than to 4.72.

EXAMPLE 3.   The number 42.645, ending in 5, is just as close to 42.64 as it is to 42.65, and we must decide which of these numbers to use in rounding off 42.645 to four significant digits. The rule in common use for this case, that is, when the digit to be dropped is 5 is as follows:†

If the digit preceding the 5 is even, it is left unchanged when the 5 is dropped; if the digit preceding the 5 is odd, it is increased by 1 when the 5 is dropped.

Thus, 42.645, rounded off to four significant digits, is 42.64.

---

* Wilczynski, E. J. and H. E. Slaught, *Plane Trigonometry and Applications*, Allyn and Bacon, New York, 1914, page 36.

† This rule is arbitrary, but leads, in general, to a smaller cumulation of errors.

EXAMPLE 4.   The number 47.365, rounded off to four significant digits, is 47.36.   The number 47.335, rounded off to four significant digits, is 47.34.

EXAMPLE 5.   The number 34.546, rounded off to three significant digits, is 34.5, since the last two figures are less than 50.   Note that by rounding off one figure at a time, the result would have been 34.6, which is incorrect.

Addition and Subtraction of Approximate Numbers

If two or more approximate numbers are added, we may assume that the absolute error of the sum will be as large as that of any of the addends. Therefore the sum should be rounded off to the same degree of precision as the least precise of the addends.

EXAMPLE 6.
$$
\begin{array}{r}
37.242 \\
9.7 \\
9.421 \\
\underline{18.20} \\
74.563
\end{array}
$$

The absolute error of the sum is 0.0005, whereas the absolute error of the addend 9.7 is 0.05.   Clearly, the error of the sum cannot be less than that of the least precise number; therefore the sum should be rounded off to 74.6.   The same rule applies to subtraction of approximate numbers.

Multiplication and Division of Approximate Numbers

It can be shown that if two or more approximate numbers are multiplied, the product can be no more accurate than the least accurate factor. Therefore the product should be rounded off to the same number of significant digits as the least accurate factor.

EXAMPLE 5. $86.72 \times 35.67 = 3093.3024$.   Since each factor has four significant figures, the product should be rounded off to 3093.   The same rule applies to the quotient of approximate numbers.

We may show that 3093 is a reasonable answer for our product by computing the limits between which the product must lie.   Taking the largest values of the measurements,

$$86.725 \times 35.675 = 3093.919375$$

Taking the smallest values of the measurements,

$$86.715 \times 35.665 = 3092.685475$$

Since the product of 86.72 and 35.67 must lie between these limits, our rounded number 3093 is a good approximation. To state the product as 3093.3024 would be indicating a degree of accuracy which clearly does not exist.

Exercise 43

COMPUTATION WITH APPROXIMATE NUMBERS

Round off each of the following numbers to 3 significant digits:

1. 2.413	**2.** 4.074	**3.** 62.32	**4.** 4,341
5. 82.49	**6.** 3.411	**7.** 47.35	**8.** 6,245
9. 8.775	**10.** 9.165	**11.** 1,076.51	**12.** 123.445
13. 1.4548	**14.** 1.4451	**15.** 1.01449	**16.** 5,632.75

Add each of the following approximate numbers:

17. 2.17 + 2.82   **18.** 13.452 + 2.36   **19.** 0.017 + 0.02 + 1.342
20. 324.7 + 0.7854 + 1.333   **21.** 643.323 + 875.2 + 2,400.32

Subtract each of the following approximate numbers:

22. 432.65 − 127.386   **23.** 13.368 − 4.65   **24.** 10.3 − 0.875
25. 0.017 − 0.002   **26.** 242.3 − 45.25   **27.** 5.00 − 0.31416

Find each of the following products. (The underlined numbers are approximate.)

28. 4.7 × 2.63   **29.** 86.5 × 28.2   **30.** 17.6 × 5
31. 2.54 × 0.0042   **32.** 210 × 3.57 × 4.21   **33.** 8700 × 4.24
34. 4.3 × 24.332   **35.** 4.3 × 24.332   **36.** 32 × 3.1416

Find each of the following quotients. (The underlined numbers are approximate.)

37. 0.0864 ÷ 5   **38.** 84.3 ÷ 36.2   **39.** 83 ÷ 0.08
40. 105.36 ÷ 22.05   **41.** 240.5 ÷ 52   **42.** 240.5 ÷ 52

In the following problems, round off answers to the proper degree of accuracy or precision:

43. What is the thickness of 72 pieces of sheet steel if each piece has a thickness of 0.024 inch?

44. A speedometer shows distances to the nearest tenth of a mile. What mileage will it show if the automobile travels 750 yards?

45. Find the area of a circle with radius 2.24 inches ($A = \pi r^2$).

46. A field of 2.45 acres produced 425 bushels of corn. What was the yield per acre?

47. A steel block 4.503 inches thick has 0.003 inch planed from one side. What is the resulting thickness?

48. The dimensions of a rectangular lot are 125.5 feet by 35.2 feet.

   (*a*) What is the perimeter of the lot?

   (*b*) What is the area of the lot?

49. How many gallons of water are there in a cylindrical water tank 58.2 inches high and with radius 16.5 inches, if there are 231 cubic inches to a gallon? ($V = \pi r^2 h$)

50. Water is flowing through a water main at the rate of 345 gallons per minute. How many gallons flow through the main in 45 minutes?

44   Zero, Fractional, and Negative Exponents

In previous sections, we studied the meaning of exponents and the rules for operating with them. The meaning assigned to an exponent, that is, a symbol indicating repeated multiplication, limited the numbers used as exponents to the set of positive integers. We now wish to extend the meaning of exponents to include fractions, negative numbers, and zero. In deciding the meaning of such expressions as

$$2^0, \ 4^{-3}, \ 9^{1/2}, \ \text{and} \ 8^{2/3}$$

we shall simply require that they obey certain basic laws of exponents. These laws are:

   1. $x^m \cdot x^n = x^{m+n}$

   2. $(x^m)^n = x^{mn}$

   3. $\dfrac{x^m}{x^n} = x^{m-n}$ $(x \neq 0)$

The Zero Exponent

Let us consider the expression x^0. What does it mean? If it is to obey Law 1, then we must have:

$$x^0 \cdot x^n = x^{0+n} = x^n \tag{1}$$

From this equality, it follows that

$$x^0 = \frac{x^n}{x^n} = 1 \qquad (x \neq 0) \tag{2}$$

We are therefore led to define the zero exponent in accordance with Equation 2, that is, for all values of x except 0,

$$x^0 = 1 \tag{3}$$

EXAMPLE 1. $2^0 = 1$

EXAMPLE 2. $(a + b - 3)^0 = 1 \qquad (a + b \neq 3)$

The Fractional Exponent $x^{1/n}$

Let us consider the expression $x^{1/2}$.   If it is to obey Law 2, we must have

$$(x^{1/2})^2 = x^{2/2} = x \tag{4}$$

It follows from statement 4 that $x^{1/2}$ is the square root of x, and we may therefore write

$$x^{1/2} = \sqrt{x} \tag{5}$$

More generally, since $(x^{1/n})^n = x^{n/n} = x$,

$$x^{1/n} = \sqrt[n]{x} \tag{6}$$

Equation 6 is therefore the definition of $x^{1/n}$.

EXAMPLE 3. $4^{1/2} = \sqrt{4} = 2$

EXAMPLE 4. $8^{1/3} = \sqrt[3]{8} = 2$

EXAMPLE 5. $(-27)^{1/3} = \sqrt[3]{-27} = -3$

The Fractional Exponent $x^{m/n}$

Let us consider the expression $x^{2/3}$.   If it is to obey Law 2, then we must have

$$(x^{2/3})^3 = x^{6/3} = x^2 \tag{7}$$

It follows from statement 7 that

$$x^{2/3} = \sqrt[3]{x^2} \tag{8}$$

On the other hand, we may write either

$$(x^{1/3})^2 = (\sqrt[3]{x})^2 \qquad \text{(by Equation 6)} \tag{9}$$

or
$$(x^{1/3})^2 = x^{2/3} = \sqrt[3]{x^2} \qquad \text{(by Equation 8)} \tag{10}$$

Therefore, by Equations 9 and 10,

$$x^{2/3} = \sqrt[3]{x^2} \qquad \text{or} \qquad (\sqrt[3]{x})^2 \tag{11}$$

More generally, since $(x^{m/n})^n = x^m$,

$$x^{m/n} = \sqrt[n]{x^m} \qquad \text{or} \qquad (\sqrt[n]{x})^m \tag{12}$$

Equation 12 is therefore the definition of $x^{m/n}$.

EXAMPLE 6. $8^{2/3} = \sqrt[3]{8^2} = \sqrt[3]{64} = 4$

or
$$8^{2/3} = (\sqrt[3]{8})^2 = (2)^2 = 4$$

EXAMPLE 7. $125^{2/3} = \sqrt[3]{125^2} = \sqrt[3]{15625} = 25$

or
$$125^{2/3} = (\sqrt[3]{125})^2 = (5)^2 = 25$$

Example 7 shows that it is advantageous to find the root first (the second equality of Equation 12), since $\sqrt[3]{125}$ is more readily found than $\sqrt[3]{15625}$.

The Negative Exponent x^{-n}

Let us consider the expression x^3/x^5. If it is to obey Law 3, we must have

$$\frac{x^3}{x^5} = x^{3-5} = x^{-2} \tag{13}$$

By the basic law of fractions, however, dividing each term of the fraction by x^3,

$$\frac{x^3}{x^5} = \frac{1}{x^2} \tag{14}$$

From Equations 13 and 14, it must follow that

$$x^{-2} = \frac{1}{x^2} \tag{15}$$

More generally,

$$x^{-n} = \frac{1}{x^n} \tag{16}$$

Equation 16 is, therefore, the definition of x^{-n}.

EXAMPLE 8. $2^{-1} = \dfrac{1}{2}$

EXAMPLE 9. $27^{-\frac{2}{3}} = \dfrac{1}{27^{\frac{2}{3}}} = \dfrac{1}{9}$

EXAMPLE 10. $\left(\dfrac{2}{3}\right)^{-2} = \dfrac{2^{-2}}{3^{-2}} = \dfrac{3^2}{2^2} = \left(\dfrac{3}{2}\right)^2 = \dfrac{9}{4}$

EXAMPLE 11. $\dfrac{1}{2^{-3}} = \dfrac{1}{\dfrac{1}{2^3}} = 2^3 = 8$

Example 10 may be generalized to obtain the relation

$$\left(\frac{a}{b}\right)^{-n} = \left(\frac{b}{a}\right)^{n} \tag{17}$$

Examples 8 and 11 show that any factor of an expression may be changed from numerator to denominator, or vice versa, by changing the sign of its exponent.

EXAMPLE 12. $\dfrac{a^{-2}b^3}{c^{-3}} = \dfrac{b^3c^3}{a^2}$

EXAMPLE 13. $\dfrac{2 \times 10^{-3}}{3 \times 10^{-4}} = \dfrac{2 \cdot 10^4}{3 \cdot 10^3} = \dfrac{2}{3} \cdot \dfrac{10^4}{10^3} = \dfrac{2}{3} \cdot 10 = \dfrac{20}{3}$

It should be noted, however, that Example 13 may be solved by a direct application of Law 3 since

$$\frac{2 \times 10^{-3}}{3 \times 10^{-4}} = \frac{2}{3} \times 10^{-3-(-4)} = \frac{2}{3} \times 10^1 = \frac{20}{3}$$

Exercise 44

ZERO, FRACTIONAL, AND NEGATIVE EXPONENTS

Evaluate each of the following expressions:

1. 2^0
5. $3a + (2b)^0$
9. $2a^0 - 3a^2 - (2a)^0$

2. $(-2)^0$
6. $3a + 2b^0$
10. $(a^2 + b^2)^0$

3. $(2a)^0$
7. $4^0 - 2^0$
11. $2^0 + 2^0$

4. $(-2a)^0$
8. $(4^0 - 2)^0$
12. $2^0 - 2^0$

Write each of the following expressions with radicals and simplify:

13. $9^{1/2}$ **14.** $8^{1/3}$ **15.** $(-8)^{1/3}$ **16.** $4^{1/2}$

17. $49^{1/2}$ **18.** $125^{1/3}$ **19.** $(-125)^{1/3}$ **20.** $(8x^3)^{1/3}$

21. $(9a^2)^{1/2}$ **22.** $32^{1/5}$ **23.** $81^{1/4}$ **24.** $(16x^4)^{1/4}$

25. $8^{2/3}$ **26.** $(-8)^{2/3}$ **27.** $4^{3/2}$ **28.** $9^{3/2}$

29. $27^{2/3}$ **30.** $27^{4/3}$ **31.** $81^{3/4}$ **32.** $(-125)^{2/3}$

33. $9^{3/2}$ **34.** $16^{3/4}$ **35.** $32^{2/5}$ **36.** $125^{2/3}$

37. $(9x^2)^{3/2}$ **38.** $(89^3)^{2/3}$ **39.** $(-8a^3)^{2/3}$ **40.** $(-27b^3)^{2/3}$

Evaluate each of the following expressions:

41. 3^{-2} **42.** 2^{-3} **43.** $\dfrac{1}{2^{-3}}$ **44.** $\dfrac{1}{3^{-2}}$

45. $\left(\dfrac{2}{3}\right)^{-3}$ **46.** $\left(\dfrac{2}{3}\right)^{-2}$ **47.** $4^{-1/2}$ **48.** $\dfrac{1}{4^{-1/2}}$

49. $(27)^{-2/3}$ **50.** $(-4)^{-3}$ **51.** $(-3)^{-2}$ **52.** $\left(\dfrac{1}{7}\right)^{-1}$

53. $\dfrac{1}{2^{-2}}$ **54.** $\dfrac{1}{9^{-1/2}}$ **55.** $\dfrac{1}{16^{-1/4}}$ **56.** $\dfrac{1}{16^{-3/4}}$

57. $\left(\dfrac{9}{16}\right)^{-1/2}$ **58.** $\left(\dfrac{4}{9}\right)^{-3/2}$ **59.** $\dfrac{1}{8^{-1/3}}$ **60.** $\dfrac{1}{8^{-2/3}}$

Write each of the following expressions with positive exponents:

61. $3a^{-1}$ **62.** $(3a)^{-1}$ **63.** $a^2b^{-1}c$ **64.** $6x^{-2}yz^{-1}$

65. $\dfrac{3}{a^{-2}b}$ **66.** $\dfrac{4a^{-1}}{b^{-2}}$ **67.** $\dfrac{1}{xy^{-2}}$ **68.** $\dfrac{2^{-1}x^2}{y^{-2}}$

69. $2ax^{-2}$ **70.** $2a^{-2}$ **71.** $\dfrac{1}{2a^{-2}}$ **72.** $\dfrac{1}{(2a)^{-2}}$

73. $ab^{-2}c$ **74.** $\dfrac{a^{-1}}{2b^{-1}}$ **75.** $\dfrac{a^{-1}}{(2b)^{-1}}$ **76.** $\left(\dfrac{a}{b}\right)^{-1}$

77. $\left(\dfrac{2}{x}\right)^{-2}$ **78.** $\left(\dfrac{4}{x^2}\right)^{-1/2}$ **79.** $\left(\dfrac{8}{x^3}\right)^{-1/3}$ **80.** $\left(\dfrac{16}{x^4}\right)^{-1/2}$

Write each of the following expressions with positive exponents. Use the basic law of fractions as in Problem 81.

81. $\dfrac{x^{-3}+x^{-2}}{x^{-2}} = \dfrac{x^{-3}+x^{-2}}{x^{-2}} \cdot \dfrac{x^3}{x^3} = \dfrac{x^0+x}{x} = \dfrac{1+x}{x}$

82. $\dfrac{2^{-1}+2^{-2}}{2^{-2}}$ **83.** $\dfrac{10^{-3}+10^{-2}}{10^2}$

84. $\dfrac{2\times 10^{-2}+6\times 10^{-3}}{4\times 10^{-2}}$ **85.** $\dfrac{a^{-1}b^{-2}+a^{-2}b^{-1}}{a^{-2}b^{-2}}$

Write each of the following expressions as a single fraction with positive exponents:

86. $a^{-1} + b^{-1}$

87. $a^{-2} + b^{-2}$

88. $2^{-2} + 2^{-3}$

89. $\dfrac{2^{-2} + 2^{-3}}{2^2}$

90. $\dfrac{2a^{-2}b - 3ab^{-2}}{a^{-1}b^{-1}}$

45 Scientific Notation

Certain measurements in scientific work are very large; other measurements are very small. For example, Avogadro's number, used in chemistry, is 606,200,000,000,000,000,000,000 and the size of a hydrogen atom is about 0.000 000 004 inch.

In order to write such numbers in a more convenient form and to indicate those digits which are significant, they are usually written in *scientific notation*. A number is written in scientific notation (sometimes called standard form) when it is written as a number between 1 and 10 multiplied by an appropriate power of 10. The number between 1 and 10 is written to show those digits, and only those digits, which are known to be significant.

EXAMPLE 1. $606,200,000,000,000,000,000,000 = 6.062 \times 10^{23}$

EXAMPLE 2. $0.000\ 000\ 004 = 4 \times 10^{-9}$

Changing from Ordinary to Scientific Notation

To write a number in scientific notation, we count the number of places the decimal point must be moved from its original position to standard position. This number is the power of 10 to be used; it is positive if the decimal point is moved to the left and negative if moved to the right. This rule may be easily remembered if we keep in mind that a positive power of 10 represents a number greater than 1, and a negative power of 10 represents a number less than 1. The reason for this procedure is clear; if we move the decimal point to the left, say three places, we divide the number by 10^3, so that to retain the original number we must multiply by 10^3. If we move the decimal point three places to the right, we multiply the number by 10^3, and must therefore multiply by 10^{-3} to retain the original value.

EXAMPLE 3. $4\,01{,}000{,}000. = 4.01 \times 10^8$

EXAMPLE 4. $0.000003\,12 = 3.12 \times 10^{-6}$

Changing from Scientific to Ordinary Notation

When a number is written in scientific notation, it is clear from the preceding discussion that the power of 10 indicates the number of places the decimal point has been moved to the left or to the right, according as the exponent is positive or negative.   We therefore move the decimal point to the right if the exponent is positive, and to the left if the exponent is negative.

EXAMPLE 5. $2.46 \times 10^4 = 24{,}600$

EXAMPLE 6. $2.32 \times 10^{-4} = 0.000232$

Using Scientific Notation in Computation

An important application of scientific notation is found in estimating the results of computations involving very small or very large numbers. This is especially true if a slide rule is used, where the location of the decimal point is best done by estimation.   Each number is rounded off to one or two significant digits and expressed in scientific notation.   The following examples will illustrate the process.

EXAMPLE 7.   Estimate the result of the following problem:

$$\frac{(0.0475)(51.95)(2.22)}{(20{,}985)(365.2)}$$

This problem may be estimated as

$$\frac{(5 \times 10^{-2})(5 \times 10^1)(2)}{(2 \times 10^4)(4 \times 10^2)} = \frac{50}{8} \times 10^{-7} = 6 \times 10^{-7}$$

The result to 3 significant digits is 6.42×10^{-7}.

EXAMPLE 8.   The sag, d, in feet, of a certain suspension bridge cable is given by

$$d = \frac{(1.42)(1100^2)}{(8)(12{,}000)}$$

This may be estimated as

$$d = \frac{(1.4)(1.1 \times 10^3)^2}{(8)(1.2 \times 10^4)}$$

$$= \frac{(1.4)(1.2 \times 10^6)}{1 \times 10^5}$$

$$= 16.8 \quad \text{(estimated value)}$$

The computed value to 3 significant digits is 17.9 feet.

Exercise 45

Scientific Notation

The following measurements occur in scientific literature. Write each of the numbers in scientific notation.

1. 357,000 miles.   The diameter of Halley's comet.
2. 30,000,000,000 centimeters per second.   The velocity of light.
3. 14,700,000 square miles.   The area of the moon.
4. 440,000,000 years.   One estimate of the age of the earth.
5. 230,000,000,000,000 horsepower.   The energy received by the earth from the sun.
6. 1,500,000 light years.   The distance to Andromeda Spiral Nebula.
7. 0.000 011.   The coefficient of linear expansion of steel.
8. 0.000 042 centimeter.   The wavelength of violet light.
9. 0.000 000 03 centimeter.   The length of the edge of a molecule.
10. 0.000 027 3 centimeter.   The photoelectric threshold wavelength of tungsten.
11. 0.000 000 000 001 60 erg.   The equivalent of one electron volt.
12. 0.00129 gram.   The weight of a cubic centimeter of air.

The following equivalents are written in scientific notation.   Change each of the numbers to ordinary notation.

13. One foot pound per minute equals 3.030×10^{-5} horsepower.
14. One foot pound per minute equals 2.260×10^{-5} kilowatts.
15. One foot pound per minute equals 2.142×10^{-5} Btu per second.
16. One Angstrom unit equals 1×10^{-8} centimeter.
17. One erg equals 2.778×10^{-14} kilowatt hour.
18. One kilowatt hour equals 3.6×10^6 joules.
19. One kilometer equals 1×10^{13} Angstrom units.

20. One kilowatt hour equals 3.67×10^{10} gram centimeters.

21. One erg equals 7.376×10^{-8} foot pound.

22. One Btu equals 1.056×10^{10} ergs.

23. One atmosphere equals 1.01325×10^6 dynes.

24. One cubic centimeter equals 3.531×10^{-5} cubic foot.

Estimate the result of each of the following computations with sufficient accuracy to locate the decimal point.

25. $\dfrac{298 \times 4.8}{64.2}$

26. $\dfrac{6.35 \times 2.58}{168}$

27. $\dfrac{1200 \times 740 \times 273}{760 \times 288}$

28. $\dfrac{372 \times 0.000234 \times 8.72}{145.3 \times 96.6 \times 0.00247}$

29. The radius in meters of the path of an alpha particle in a certain cyclotron is given by the formula

$$r = \frac{6.71 \times 10^{-27} \times 10^5}{5 \times 3.2 \times 10^{-19}}$$

Estimate the value of r.

30. The perfect efficiency of a certain inclined plane is given by the formula

$$E = \frac{210 \times 2.5}{64 \times 15}$$

Estimate the value of E.

31. The horsepower transmitted by a certain belt is given by the formula

$$P = \frac{62.5 \times 1400\pi}{33,000}$$

Estimate the value of P.

32. The horsepower required to pump water at a certain rate is given by the formula

$$P = \frac{160 \times 62.4 \times 25}{0.60 \times 33,000}$$

Estimate the value of P.

33. The electrical current in a certain coil is given by the formula

$$I = \frac{20 \times 16 \times 0.18 \times 1.28}{2\pi \times 30}$$

Estimate the value of I.

CHAPTER 9

Logarithms

An important application of exponents deals with the simplification of numerical computations, in particular, the computations that involve multiplication, division, powers, and roots. Exponents used for this purpose are called *logarithms*. Although at the present time modern computing machines have decreased our dependence on logarithmic computation, there are still many situations in which such computation is our only resource. Aside from their use for numerical calculations, logarithms play an important role in the theory and applications of higher mathematics.

Generalization of the Concept of Exponents

It is instructive to review the generalization of the concept of exponents which leads to their use for computation. This will serve both to summarize certain principles of exponents to which we shall allude, and to illustrate the kind of thinking which is basic to the development of a mathematical system.

Our first definition of exponent was restricted to the set of positive integers. Operating within this set, the exponent was defined as a symbol denoting the number of times a given quantity is to be used as a factor. Thus

$$x^m = x \cdot x \cdot x \ldots \text{ (a product of } m \text{ factors, } m \text{ a positive integer)} \quad (1)$$

From this definition, the following theorems were derived directly:

$$x^m \cdot x^n = x^{m+n} \tag{2}$$

$$\frac{x^m}{x^n} = x^{m-n} \; (m > n, \, x \neq 0) \qquad \text{or} \qquad \frac{x^m}{x^n} = \frac{1}{x^{n-m}} \; (n > m, \, x \neq 0) \tag{3}$$

$$(x^m)^n = x^{mn} \tag{4}$$

These relations, at this point, were restricted to positive, integral values of m and n.

Our first generalization of these laws was made in order to extend the admissible values of exponents to negative integers and zero. Now nothing really compels us to introduce such symbols as x^0 or x^{-3} except our desire to extend our knowledge to include more extensive sets of elements. We found the following interpretation of the negative and zero exponent was necessary to satisfy the relations 2, 3, and 4:

$$x^0 = 1 \qquad (x \neq 0) \tag{5}$$

$$x^{-n} = \frac{1}{x^n} \qquad (x \neq 0) \tag{6}$$

The domain of an exponent was thus extended to the entire set of integers. (Note that 3 can now be written $\frac{x^m}{x^n} = x^{m-n}$ without reference to the relative size of m and n.)

Our next extension was made to include the set of all rational numbers. Again, to maintain the consistency of the laws of exponents, it was necessary to assign certain definitions to rational exponents.

$$x^{1/q} = \sqrt[q]{x} \tag{7}$$

$$x^{p/q} = \sqrt[q]{x^p} = (\sqrt[q]{x})^p * \tag{8}$$

Although it is beyond the scope of this book, it can be shown that the domain of an exponent can be further extended to include irrational numbers. As a matter of fact, this last extension is quite necessary to the study of logarithms since most logarithms are irrational exponents.

Thus the initial definition of an exponent as a symbol denoting repeated multiplication does not have meaning for all real numbers because it is absurd to speak, for example, of the product of a fractional number of factors. However, the exponent symbol is defined for all values in the set of real numbers in such a way as to satisfy the stated laws of exponents, and in this generalized form the definitions are not only logically admissible but are very useful as well.

* We must exclude the case where $x < 0$ and q is even.

It is appropriate to mention here that, from a mathematician's point of view, the primary aim of such generalization is not its usefulness—although this may be a welcome by-product.   Fundamentally the mathematician is seeking to maintain the consistency of a logical structure while following a compelling curiosity concerning the effects of generalization.

Definition of a Logarithm

Using the extended concept that an exponent may be any real number, we can now express every positive real number N in the form of an exponential quantity b^x.   When this is done, the exponent x is called the logarithm of N to the base b.   In symbols, these relations are written:

$$\text{If} \qquad N = b^x \qquad\qquad (9)$$
$$\text{then} \quad \log_b N = x \qquad\qquad (10)$$

Statement 10 is read "The logarithm (or log, for short) of N to the base b is x."

For example, the number 64 can be written as 4^3.   Therefore the logarithm of 64 to the base 4 is 3.   Note that a logarithm is an exponent and that the statements

$$64 = 4^3 \qquad\qquad (11)$$
$$\log_4 64 = 3 \qquad\qquad (12)$$

express the same relationship in two ways.   The first is called an *exponential form* and the second a *logarithmic form* of the relationship.

Below are written several exponential statements followed by their corresponding logarithmic forms and the verbal statement of the logarithmic relationship:

(a) $7^2 = 49$ $\qquad \log_7 49 = 2$ $\qquad$ The logarithm of 49 to the base 7 is 2.

(b) $5^0 = 1$ $\qquad \log_5 1 = 0$ $\qquad$ The logarithm of 1 to the base 5 is 0.

(c) $10^{-1} = \frac{1}{10}$ $\qquad \log_{10} \frac{1}{10} = -1$ $\qquad$ The logarithm of $\frac{1}{10}$ to the base 10 is -1.

(d) $9^{1/2} = 3$ $\qquad \log_9 3 = \frac{1}{2}$ $\qquad$ The logarithm of 3 to the base 9 is $\frac{1}{2}$.

(e) $8^{2/3} = 4$ $\qquad \log_8 4 = \frac{2}{3}$ $\qquad$ The logarithm of 4 to the base 8 is $\frac{2}{3}$.

As a result of these considerations, the following formal definition of a logarithm should now have meaning for the student:

The logarithm of a positive real number N to a given positive base b ($b \neq 1$) is the exponent x to which the base must be raised to equal the number.

The examples which follow illustrate methods of determing the value of N, b, or x in this relationship.

EXAMPLE 1.   Find the value of x if $\log_2 8 = x$.
The equivalent exponential form is $2^x = 8$.   Therefore $x = 3$.

EXAMPLE 2.   Find the value of N if $\log_9 N = \frac{3}{2}$.
The equivalent exponential form is $9^{3/2} = N$.   Therefore $N = (\sqrt{9})^3 = 27$.

EXAMPLE 3.   What is the value of b in the relation $\log_b \frac{1}{16} = -2$?
The equivalent exponential form is $b^{-2} = \frac{1}{16}$.   Then $1/b^2 = \frac{1}{16}$ and $b = \pm 4$.   Since the base b must be a positive number, $b = 4$

Exercise 46

THE MEANING OF A LOGARITHM

Express each of the following in logarithmic form:

1. $9^2 = 81$	**2.** $10^3 = 1000$	**3.** $2^4 = 16$
4. $64^{1/3} = 4$	**5.** $64^{1/2} = 8$	**6.** $3^{-2} = \frac{1}{9}$
7. $6^0 = 1$	**8.** $32^{1/5} = 2$	**9.** $5^3 = 125$
10. $4^{3/2} = 8$	**11.** $2^{-3} = \frac{1}{8}$	**12.** $81^{3/4} = 27$

Express each of the following in exponential form:

13. $\log_5 25 = 2$	**14.** $\log_4 64 = 3$	**15.** $\log_{16} 4 = \frac{1}{2}$
16. $\log_8 2 = \frac{1}{3}$	**17.** $\log_5 \frac{1}{25} = -2$	**18.** $\log_{10} 0.01 = -2$
19. $\log_9 1 = 0$	**20.** $\log_{16} 8 = \frac{3}{4}$	**21.** $\log_4 32 = \frac{5}{2}$
22. $\log_{10} 0.1 = -1$	**23.** $\log_8 8 = 1$	**24.** $\log_5 625 = 4$

Find the value of each of the given logarithms:

25. $\log_3 27$	**26.** $\log_2 64$	**27.** $\log_{10} 10,000$
28. $\log_{10} 1$	**29.** $\log_3 \frac{1}{3}$	**30.** $\log_4 4$
31. $\log_8 16$	**32.** $\log_7 (7^4)$	**33.** $\log_3 \sqrt{3}$

Determine the value of the literal quantity in each of the following:

34. $\log_b 36 = 2$ **35.** $\log_8 N = -2$ **36.** $\log_6 216 = x$

37. $\log_{10} N = 1$ **38.** $\log_{49} 7 = x$ **39.** $\log_b 32 = \frac{5}{4}$

40. $\log_b \frac{1}{5} = -1$ **41.** $\log_{10} N = -3$ **42.** $\log_{64} N = \frac{2}{3}$

Find the value of each logarithm. How is the logarithm in part c related to the logarithms in parts a and b?

43. (a) $\log_2 8$ (b) $\log_2 4$ (c) $\log_2 (8 \cdot 4)$

44. (a) $\log_3 27$ (b) $\log_3 9$ (c) $\log_3 (27 \cdot 9)$

45. (a) $\log_{10} \frac{1}{10}$ (b) $\log_{10} 1000$ (c) $\log_{10} (\frac{1}{10} \cdot 1000)$

46. (a) $\log_{10} 10$ (b) $\log_{10} \frac{1}{100}$ (c) $\log_{10} (10 \cdot \frac{1}{100})$

Find the value of each logarithm. How is the logarithm in part c related to the logarithms in parts a and b?

47. (a) $\log_2 32$ (b) $\log_2 2$ (c) $\log_2 (\frac{32}{2})$

48. (a) $\log_5 25$ (b) $\log_5 125$ (c) $\log_5 (\frac{25}{125})$

49. (a) $\log_{10} 1000$ (b) $\log_{10} 100$ (c) $\log_{10} (\frac{1000}{100})$

50. (a) $\log_{10} \frac{1}{10}$ (b) $\log_{10} \frac{1}{100}$ (c) $\log_{10} (\frac{1}{10} \div \frac{1}{100})$

51. Find the value of (a) $\log_2 1$, (b) $\log_5 1$, (c) $\log_b 1$.

52. Find the value of (a) $\log_{10} 10$, (b) $\log_3 3$, (c) $\log_b b$.

53. Can the logarithm of a positive number be negative? If so, give an example. If not, explain why not.

54. Using a positive base, can a negative number have a logarithm in the set of real numbers? If so, give an example. If not, explain why not.

47 The Laws of Logarithms

By definition, logarithms are exponents. Therefore the laws which govern operations with exponents determine analogous laws which apply to operations with logarithms. In particular, the following laws of exponents:

$$b^x \cdot b^y = b^{x+y} \tag{1}$$

$$\frac{b^x}{b^y} = b^{x-y} \qquad (b \neq 0) \tag{2}$$

$$(b^x)^y = b^{xy} \tag{3}$$

lead to statements concerning logarithms which are of sufficient importance to warrant development and proof.

Logarithm of a Product

A numerical example may suggest that a simple relationship exists between the logarithm of a product and the logarithms of its factors. To illustrate, consider the logarithms

$$\log_2 (8)(32) = \log_2 256 = 8$$
$$\log_2 8 = 3$$
$$\log_2 32 = 5$$

In this example, the logarithm of the product equals the sum of the logarithms of its factors. Of course, this one example, or many such examples, do not prove that this relationship is always true.   To establish it for all admissible values of the variables, we shall prove Theorem 1.

Theorem 1. $\log_b MN = \log_b M + \log_b N$ $\hspace{2cm}$ (4)

Proof.   Let $x = \log_b M$ and $y = \log_b N$
$\hspace{1.5cm}$ In exponential form, $M = b^x$ and $N = b^y$
$\hspace{1.5cm}$ Multiplying M by N, $MN = b^x \cdot b^y = b^{x+y}$
$\hspace{1.5cm}$ In logarithmic form, $\log_b MN = x + y$
$\hspace{1.5cm}$ Therefore, by substitution, $\log_b MN = \log_b M + \log_b N$

Logarithm of a Quotient

Note the relationship of the logarithms in the following:

$$\log_3 243 = 5$$
$$\log_3 81 = 4$$
$$\log_3 \tfrac{243}{81} = \log_3 3 = 1$$

It is evident that in this instance, the logarithm of the quotient is equal to the logarithm of the numerator minus the logarithm of the denominator.   We now prove the following theorem for all admissible values of the variables.

Theorem 2. $\log_b \dfrac{M}{N} = \log_b M - \log_b N$ $\hspace{2cm}$ (5)

Proof.   Let $x = \log_b M$ and $y = \log_b N$
$\hspace{1.5cm}$ In exponential form, $M = b^x$ and $N = b^y$
$\hspace{1.5cm}$ Dividing M by N, $\dfrac{M}{N} = \dfrac{b^x}{b^y} = b^{x-y}$
$\hspace{1.5cm}$ In logarithmic form, $\log_b \dfrac{M}{N} = x - y$
$\hspace{1.5cm}$ Therefore, by substitution, $\log_b \dfrac{M}{N} = \log_b M - \log_b N$

Logarithm of a Power

Consider the following statements:

$$\log_2 64 = 6$$
$$\log_2 \sqrt[3]{64} = \log_2 64^{1/3} = \log_2 4 = 2$$

Thus the logarithm of the $\frac{1}{3}$ power of 64 is equal to $\frac{1}{3}$ of the logarithm of 64. In general, for all values of n in the set of real numbers we prove the following theorem.

Theorem 3. $\log_b M^n = n \log_b M$ \hfill (6)

Proof.   Let $x = \log_b M$
In exponential form, $M = b^x$
Raising to the nth power, $M^n = (b^x)^n = b^{nx}$
In logarithmic form, $\log_b M^n = nx$
Therefore, by substitution, $\log_b M^n = n \log_b M$

Laws of Logarithms

The relationships 4, 5, and 6, known as laws of logarithms, are used repeatedly in the application of logarithms to problems of computation. Their usefulness will be increased if the student learns to express them in words as well as in their symbolic form.

Law 1.   The logarithm of a product is equal to the sum of the logarithms of its factors.   That is, $\log_b MN = \log_b M + \log_b N$.

Law 2.   The logarithm of a quotient is equal to the logarithm of its numerator minus the logarithm of its denominator.   That is, $\log_b M/N = \log_b M - \log_b N$.

Law 3.   The Logarithm of a power of a number is equal to the exponent times the logarithm of the number.   That is, $\log_b M^n = n \log_b M$.

EXAMPLE 1.   Given $\log_{10} 3 = 0.4771$, find $\log_{10} 81$.

$$\log_{10} 81 = \log_{10} 3^4$$
$$= 4 \log_{10} 3 \qquad \text{(by Law 3)}$$
$$= 4(0.4771) = 1.9084$$

EXAMPLE 2.   Given $\log_{10} 2 = 0.3010$ and $\log_{10} 3 = 0.4771$, find $\log_{10} \sqrt{6}$.

$$\log_{10} \sqrt{6} = \log_{10} (2 \cdot 3)^{\frac{1}{2}}$$

$$= \tfrac{1}{2} \log_{10} (2)(3) \qquad \text{(by Law 3)}$$

$$= \tfrac{1}{2}(\log_{10} 2 + \log_{10} 3) \qquad \text{(by Law 1)}$$

$$= \tfrac{1}{2}(0.3010 + 0.4771) = 0.3890$$

EXAMPLE 3.   If $\log_3 5 = a$, find $\log_3 \tfrac{1}{5}$.

$$\log_3 \tfrac{1}{5} = \log_3 1 - \log_3 5 \qquad \text{(by Law 2)}$$

$$= 0 - a = -a$$

EXAMPLE 4.   Express $\log_{10} \pi + 2 \log_{10} r$ as a single logarithm.

$$\log_{10} \pi + 2 \log_{10} r = \log_{10} \pi + \log_{10} r^2 \qquad \text{(by Law 3)}$$

$$= \log_{10} \pi r^2 \qquad \text{(by Law 1)}$$

Exercise 47

LAWS OF LOGARITHMS

If $a = \log_5 x$ and $b = \log_5 y$, express each of the following in terms of a and b:

1. $\log_5 xy$

2. $\log_5 \dfrac{x}{y}$

3. $\log_5 \dfrac{y}{x}$

4. $\log_5 x^5$

5. $\log_5 \sqrt{y}$

6. $\log_5 x^2 y$

7. $\log_5 \sqrt[3]{x/y}$

8. $\log_5 x^{\frac{2}{3}} y^3$

9. $\log_5 \dfrac{\sqrt{x}}{y^2}$

10. $\log_5 \sqrt{xy}$

11. $\log_5 \dfrac{1}{x}$

12. $\log_5 \dfrac{1}{xy}$

If $\log_{10} 2 = 0.3010$, $\log_{10} 3 = 0.4771$, and $\log_{10} 5 = 0.6990$, find the logarithms of each of the following quantities to the base 10.   Express each result to four decimal places.

13. 4

14. $\sqrt{3}$

15. 30

16. $\tfrac{3}{2}$

17. 6^7

18. 18

19. $\tfrac{25}{3}$,

20. $\sqrt[3]{15}$

21. $\sqrt{75}$

22. $\tfrac{9}{4}$

23. $\sqrt[5]{10}$

24. $12^{\frac{3}{4}}$.

If $\log_{10} 30.2 = 1.4800$ and $\log_{10} 8.83 = 0.9460$, find the following logarithms to four decimal places:

25. $\log_{10} \sqrt[3]{8.83}$

26. $\log_{10} (30.2)^4$

27. $\log_{10} \dfrac{30.2}{8.83}$

28. $\log_{10} (30.2)(8.83)$

29. $\log_{10} \dfrac{(8.83)^2}{\sqrt{30.2}}$

30. $\log_{10} 8.83 \sqrt[5]{30.2}$

Express each of the given expressions as a single logarithm whose coefficient is 1:

31. $4 \log_3 x$ **32.** $\log_2 y - \log_2 z$

33. $3 \log_{10} U + 2 \log_{10} V$ **34.** $\log_b p + \log_b q - \log_b r$

35. $\frac{1}{3} \log_5 x - 2 \log_5 y$ **36.** $\log_{10} 4 - \log_{10} 3 + \log_{10} \pi + 3 \log_{10} r$

Express $\log x$ in terms of $\log a$ and $\log b$.   (All logarithms are to the same base.)

37. $x = ab$ **38.** $x = a^2 b$ **39.** $x = a^3 b^2$

40. $x = \dfrac{a}{b}$ **41.** $x = \dfrac{b}{a}$ **42.** $x = \sqrt{\dfrac{a}{b}}$

43. $x = \dfrac{\sqrt{a}}{b}$ **44.** $x = \left(\dfrac{b}{a}\right)^2$ **45.** $x = \dfrac{\sqrt[3]{a}}{\sqrt{b}}$

46. $x = (ab)^3$ **47.** $x = a^2 \sqrt{b}$ **48.** $x = (\sqrt{ab})^3$

In each of the following, express $\log y$ as an algebraic sum of logarithms, as illustrated in Problem 49, and prove each statement by using Laws 1, 2, and 3 of this section.   (All logarithms are to the same base.)

49. $y = \pi r^2 h$ **50.** $y = \sqrt{h/16}$ **51.** $y = prt$
     $\log y = \log \pi + 2 \log r + \log h$

52. $y = 4\pi r^2$ **53.** $y = p(1.06)^n$ **54.** $y = KT/p$

48   Common Logarithms

Systems of Logarithms

Logarithms are said to be in the same system if they have the same base.   Two systems of logarithms are in general use.   In one of these, the base is an irrational number symbolized by e, whose value is approximately 2.718.   In the other, the base is the number 10.   Logarithms to the base e, called *natural logarithms*, have certain advantages in the theory and solution of problems in advanced mathematics.   Logarithms to the base 10, called *common logarithms*, are convenient for simplifying ordinary numerical computations.

Our study will be confined to the system of common logarithms.   Since the base 10 will be used so frequently, hereafter, if no reference is made to the base of a logarithm, the base 10 will be understood.   Thus the logarithm of N means the common logarithm of N, and

$$\log N \text{ means } \log_{10} N$$

The Two Parts of a Logarithm

The logarithm of 10,000 is 4 since $10^4 = 10,000$.   Similarly, the logarithms of all other integral powers of 10 are integers.   For example,

$$\log 1000 = 3 \qquad \log 0.1 = -1$$
$$\log 100 = 2 \qquad \log 0.01 = -2$$
$$\log 10 = 1 \qquad \log 0.001 = -3$$
$$\log 1 = 0 \qquad \log 0.0001 = -4$$

This listing suggests that the logarithm of a number between 100 and 1000 will be greater than 2 and less than 3, the logarithm of a number between 10 and 100 will be greater than 1 and less than 2, etc.   Therefore the logarithm of, say, 695 is $+2$ followed by a decimal fraction, and, in general, a logarithm consists of two parts—an integer and a decimal fraction.   The integral part of the logarithm is called its *characteristic*, and the decimal part is called its *mantissa*.

Let us assume that the logarithm of the number 695 is 2.842.   Then $+2$ is the characteristic of this logarithm, and $+0.842$ is the mantissa. If we now list the logarithms of 695 times an integral power of 10, an important property of characteristics and mantissas can be observed. Thus

$\log 695 = 2.842$

$\log 69.5 = \log(0.1 \times 695) = \log 0.1 + \log 695 = -1 + 2.842 = 1.842$

$\log 6.95 = \log(0.01 \times 695) = \log 0.01 + \log 695 = -2 + 2.842$
$$= 0.842$$

$\log 0.695 = \log(0.001 \times 695) = \log 0.001 + \log 695 = -3 + 2.842$
$$= 0.842 - 1$$

$\log 0.0695 = \log(0.0001 \times 695) = \log 0.0001 + \log 695 = -4 + 2.842$
$$= 0.842 - 2$$

It can be seen that if we choose to keep the decimal part of the logarithm positive, the mantissa is constant for the given sequence of significant figures.   In other words, the value of the mantissa is not affected by the position of the decimal point in the sequence 695.   Furthermore, the characteristic of the logarithm changes in a regular pattern according to the position of the decimal point in the sequence.   This is an illustration of the important property that the sequence of significant digits in a given number may be used to determine the mantissa of its logarithm while the position of the decimal point in the sequence may be used to determine the characteristic of its logarithm.

Note, further, the convention introduced here for writing a logarithm so that the mantissa is always positive. When the characteristic is positive, it is combined with the mantissa to form a single number, as

$$\log 69.5 = 1.842$$

but when the characteristic is negative, it is written after the mantissa to form a binomial expression, as

$$\log 0.0695 = 0.842 - 2$$

Determining the Characteristic of a Logarithm

It will be recalled (Section 45) that every number N can be written in scientific notation in the form

$$N = c \cdot 10^k$$

where k is an integer and $1 < c < 10$. It follows that

$$\log N = \log(c \cdot 10^k)$$
$$\log N = \log c + k \log 10$$
$$\log N = \log c + k$$

Since $1 < c < 10$ and k is an integer, $\log c$ represents the mantissa of $\log N$ and k represents its characteristic. This fact may be utilized to determine the characteristic of the logarithm of a given number as follows:

Write the number in scientific notation. The characteristic of its logarithm is numerically equal to the power of 10.

This relationship is illustrated in the following table.

TABLE 1

Number N	N Expressed in Scientific Notation	Characteristic of $\log N$	Logarithm
6950	6.95×10^3	3	3.842
69.5	6.95×10^1	1	1.842
6.95	6.95×10^0	0	0.842
0.0695	6.95×10^{-2}	−2	0.842 − 2
0.000695	6.95×10^{-4}	−4	0.842 − 4

Determining the Mantissa of a Logarithm

Earlier, we made the assumption

$$\log 695 = 2.842$$

The mantissa of this logarithm can be computed to any required number of significant figures by methods used in advanced mathematics. Since such computations are difficult and tedious, the results are tabulated for convenient reference. A table of mantissas approximated to four decimal places is given in Table II. This table gives directly the mantissa of the logarithm of any number which has no more than three significant figures. In Section 50, we learn how this table may be extended with reasonable accuracy to obtain the logarithm of a number with four significant figures.

To find the mantissa of the logarithm of 695 in Table II, locate the first two significant figures in the column headed by N. The third significant figure is found as a column heading across the top of the table. The required mantissa is found in the row beginning with 69 and in the column headed by 5. This entry is 8420 and represents the mantissa of the logarithm of 695 correct to four decimal places. Since virtually all entries in the table are approximate numbers, the fact that a given logarithm is approximate is assumed without further notation. Therefore we write

$$\log 695 = 2.8420$$

EXAMPLE 1.   Find the logarithm of 30,700.

$30,700 = 3.07 \times 10^4$. The characteristic is 4. To find the mantissa, locate the entry in Table II, which is in the row beginning with 30, and in the column headed by 7. It is 4871. Therefore $\log 30,700 = 4.4871$.

EXAMPLE 2.   What is the value of log 0.0148?

$0.0148 = 1.48 \times 10^{-2}$. The characteristic is -2. The mantissa of the sequence of digits 148 is found in Table II to be 1703. Therefore $\log 0.0148 = 0.1703 - 2$.

Antilogarithms

If $\log N = x$, then N is said to be the *antilogarithm* (or antilog) of x. That is, N is the number corresponding to a given logarithm x. For example,

if $\log 695 = 2.842$
then   antilog $2.842 = 695$

Since the value of a logarithm is found by using a table of logarithms, an antilogarithm is evaluated by using the same table in reverse order.

It should be remembered that a table of logarithms lists only the mantissas of the logarithms. Therefore to find the antilogarithm of 2.842, we locate the mantissa 0.842 in the body of the table, and note that it corresponds to the sequence of significant digits 695. Since the characteristic is $+2$, the required number is

$$6.95 \times 10^2 = 695$$

EXAMPLE 3.   Find the antilogarithm of 1.4349 to three significant figures.

The mantissa 0.4349 does not appear in Table II. However, it falls between the two listed mantissas:

0.4346, whose antilogarithm is 272
0.4362, whose antilogarithm is 273

Since the given mantissa is closer in value to the mantissa 0.4346, the three-figure sequence of digits in the antilogarithm is 272. Using the characteristic $+1$, we have

$$2.72 \times 10^1 = 27.2$$

EXAMPLE 4.   Find N if $\log N = 0.7917 - 3$.

Using Table II, the mantissa 0.7917 is found in the row which begins with 61 and in the column headed by 9. Therefore the antilogarithm is 619. Since the characteristic is -3, we have

$$6.19 \times 10^{-3} = 0.00619$$

Exercise 48

COMMON LOGARITHMS

Find the characteristic of the logarithm of each of the following numbers by writing each in scientific notation:

1. 4670	**2.** 5.83	**3.** 67.2
4. 0.867	**5.** 50,800	**6.** 0.00713
7. 12	**8.** 0.0102	**9.** 2
10. 186,000	**11.** 0.0003	**12.** 9,200
13. 0.038	**14.** 4000	**15.** 0.00592

Given the sequence of significant figures 409, place the decimal point correctly in this sequence if the characteristic of its logarithm is:

16. +3   **17.** +5   **18.** 0

19. −2   **20.** +1   **21.** −4

Use Table II to find the logarithm of each of the following numbers. Be sure to indicate both characteristic and mantissa.

22. 287   **23.** 0.0821   **24.** 49,000

25. 10.6   **26.** 0.609   **27.** 0.007

28. 8   **29.** 3.07   **30.** 4,950

31. 0.093   **32.** 3,740,000   **33.** 20,100

34. 1.91   **35.** 97.6   **36.** 0.0000854

37. 10,000   **38.** 2.83×10^7 **39.** 4.07×10^{-6}

Use Table II to find the antilogarithms of each of the following logarithms correct to three significant figures.

40. 0.9571   **41.** 0.6149   **42.** 1.7716

43. 3.8780   **44.** 0.4232 − 2   **45.** 0.1303 − 1

46. 2.9200   **47.** 2.0000   **48.** 0.6167 − 3

49. 0.7367 − 1   **50.** 0.7045   **51.** 0.4000 − 6

52. 5.5180   **53.** 0.8804 − 4   **54.** 0.0400 − 6

Given $\log x = 2.8882$ and $\log y = 1.6021$, find the antilogarithm of:

55. $\log x + \log y$ **56.** $2 \log x$ **57.** $3 \log y$

58. $\log x - \log y$ **59.** $\frac{1}{2} \log x$ **60.** $\frac{1}{3} \log x$

61. $\frac{1}{2} \log y$ **62.** $\frac{1}{3} \log y$ **63.** $2 \log x + \log y$

64. $\log x + 2 \log y$ **65.** $3 \log x - \frac{1}{2} \log y$ **66.** $\log x^3$

49 Logarithmic Computations

We are now ready to apply logarithms to the computation of products, quotients, powers, and roots. Basically, just two steps are involved in determining the value of an expression which consists of a combination of these stated operations:

1. The logarithm of the entire expression is obtained by applying the three laws of logarithms (Section 47).

2. The antilogarithm then determines the value of the given expression.

Although the theory of such logarithmic computation is simple, there are certain details of procedure that warrant careful illustration. In particular, the requirements of maintaining a positive mantissa and an integral characteristic give rise to a device which is frequently useful in computing quotients or products. Furthermore, some systematic method of outlining the details of the entire computation is strongly recommended as an aid to speed and accuracy. These considerations are illustrated in the examples which follow.

Multiplication

The simplification of a product by means of logarithms is based on Law 1 (Section 47) which is restated for reference:

The logarithm of a product is equal to the sum of the logarithms of its factors.

EXAMPLE 1. Compute the value of N if $N = 36.9 \times 0.875 \times 4.26$.

By Law 1, we have $\log N = \log 36.9 + \log 0.875 + \log 4.26$.

$$
\begin{aligned}
\log 36.9 \ &= 1.5670 \\
\log 0.875 &= 0.9420 - 1 \\
\log 4.26 \ &= 0.6294 \\
\hline
\log N \quad\ &= \overline{3.1384 - 1} = 2.1384 \\
N = \text{antilog } 2.1384 &= 1.37 \times 10^2 \\
N &= 137
\end{aligned}
$$

Division

Law 2 of the laws of logarithms, used to simplify a quotient, stated:

The logarithm of a quotient is equal to the logarithm of its numerator minus the logarithm of its denominator.

If the mantissa and characteristic are each smaller in the denominator than in the numerator, the subtraction is performed easily; if this is not so, it becomes necessary, or at least highly desirable, to rewrite one of the logarithms in an equivalent form.

EXAMPLE 2. Evaluate $\dfrac{0.237}{58.1}$.

Let $Q = (0.237) \div (58.1)$
Then log $Q = $ log $0.237 - $ log 58.1 (by Law 2)
log $0.237 = 0.3747 - 1$
log 58.1 $= \underline{1.7642}$

This subtraction is an awkward one to perform if we are to maintain a positive mantissa, but it is greatly simplified if 2 is added to and subtracted from the first logarithm.   We then have

log $0.237 = 2.3747 - 3$
log 58.1 $= 1.7642$
log Q $= \overline{0.6105} - 3$
$Q = $ antilog $0.6105 - 3 = 4.08 \times 10^{-3}$
$Q = 0.00408$

Powers and Roots

To calculate a given power of a number, reference is made to Law 3 of the laws of logarithms:

The logarithm of a power of a number is equal to the exponent times the logarithm of the number.

Since an indicated root of a number is expressible as a fractional power, this rule is the basis for computing roots as well as powers.   When the exponent is fractional, it may be necessary to change the form of the logarithm to maintain an integral characteristic.

EXAMPLE 3.   Compute the cube root of 0.0534.

$$\text{Let } R = \sqrt[3]{0.0534} = (0.0534)^{1/3}$$
$$\text{Then } \log R = \frac{1}{3} \log 0.0534 \qquad \text{(by Law 3)}$$
$$= \frac{1}{3} (0.7275 - 2)$$

Before dividing the logarithm by 3, it should be rewritten in an equivalent form so that the negative characteristic is a multiple of 3.   Therefore we write

$$\log R = \frac{1}{3} (1.7275 - 3)$$
$$= 0.5758 - 1$$
$$R = \text{antilog}(0.5758 - 1) = 3.77 \times 10^{-1}$$
$$R = 0.377$$

General Computations

It is possible to calculate a great variety of numerical expressions by means of logarithms. If a computation involves sums or differences, however, the operations of addition and subtraction must be performed arithmetically after the value of each term is obtained.

EXAMPLE 4.   Evaluate $\sqrt{\dfrac{x^3 + 1}{x^3 - 1}}$ where $x = 1.47$.

$\log x = \log 1.47 = 0.1673$

$\log x^3 = 3 \log x = 0.5019$ (by Law 3)

$x^3 = \text{antilog } 0.5019 = 3.18$

Then $x^3 + 1 = 4.18;\ x^3 - 1 = 2.18$

Let $N = \sqrt{\dfrac{x^3 + 1}{x^3 - 1}} = \sqrt{\dfrac{4.18}{2.18}}$

$\log N = \frac{1}{2}(\log 4.18 - \log 2.18)$ (by Laws 2, 3)

$\quad\quad = \frac{1}{2}(0.6212 - 0.3385)$

$\quad\quad = \frac{1}{2}(0.2827)$

$\quad\quad = 0.1414$

$N = \text{antilog } 0.1414 = 1.38$

EXAMPLE 5.   Find the value of the expression $\dfrac{(\sqrt{0.0139})(0.481)}{2.86}$.

Let N be the result.   Then

$\log N = \frac{1}{2} \log 0.0139 + \log 0.481 - \log 2.86$

$\log 0.0139 = 0.1430 - 2$

$\frac{1}{2} \log 0.0139 = 0.0715 - 1$

$\log 0.481 = 0.6821 - 1$

$\frac{1}{2} \log 0.0139 + \log 0.481 = \overline{0.7536 - 2}$

$\log 2.86 = 0.4564$

$\log N = \overline{0.2972 - 2}$

$N = 0.0198$

Exercise 49

LOGARITHMIC COMPUTATION

Evaluate each of the following by use of a table of logarithms.   Express results to three significant figures.

1. 36.7×9.08

2. $\dfrac{8.94}{0.474}$

3. $\sqrt{83.6}$

4. $\sqrt[3]{0.212}$

5. $(0.0813)^4$

6. $(0.0135)(0.00284)$

7. $\dfrac{0.0509}{247}$

8. $\sqrt[3]{7.86}$

9. $\sqrt[4]{0.0136}$

10. $(1.03)^{10}$

11. $(8740)(0.634)$

12. $\sqrt[5]{513}$

13. $\dfrac{3.14}{46.5}$

14. $(41.9)^3$

15. $(0.455)^{2\frac{3}{5}}$

16. $0.841(0.0265)^2$

17. $\dfrac{(13.6)^3}{64,700}$

18. $51.8\sqrt{5.05}$

19. $\dfrac{\sqrt{0.000872}}{0.0196}$

20. $(28.4 \times 1.58)^3$

21. $\sqrt{\dfrac{82.2}{0.931}}$

22. $\sqrt[3]{\dfrac{37.1}{(8.52)(0.753)}}$

23. $\sqrt[5]{\dfrac{1}{3.43}}$

24. $(1.42)^{0.7}$

25. $(2.83 \times 10^{-6})^3$

26. $\left(\dfrac{3.87}{51.4}\right)^3$

27. $\dfrac{\sqrt[4]{679} \times 0.876}{(16.6)^3}$

28. $0.0719\sqrt[5]{\dfrac{0.463}{2.28}}$

29. $\dfrac{(48.7)(3.83)}{(2.93)(0.0419)}$

30. $\sqrt[3]{(9.16 \times 10^{-5})^2}$

31. $\dfrac{39.7(4.29)^2}{\sqrt{88.6}}$

Given $r = 6.17$, $h = 9.32$, and $\pi = 3.14$, find the value of each of the following expressions from geometry:

32. $\frac{4}{3}\pi r^3$

33. $4\pi r^2$

34. $\frac{1}{3}\pi r^2 h$

35. $2\pi r h$

36. $\frac{2}{3}\pi r^3 + \pi r^2$

37. $\pi r\sqrt{h^2 + r^2}$

38. $\frac{1}{3}\pi h(3r^2 - h^2)$

39. $2\pi r(h + r)$

40. $\dfrac{\pi h^2}{3}(3r - h)$

If a principal of P dollars is invested at a compound interest rate of r per cent per period, the amount at the end of n periods will be

$$A = P(1 + r)^n$$

Using this formula, find approximately how much \$100 will amount to if:

41. $n = 10, r = 3\%$

42. $n = 20, r = 3\%$

43. $n = 10, r = 6\%$

44. $n = 20, r = 1.5\%$

45. $n = 30, r = 5\%$

46. $n = 100, r = 1\%$

47. The approximate number of seconds, t, required for an object to strike the ground when dropped from an initial height of h feet can be calculated from the formula $t = \sqrt{h/16.1}$.   Determine the time required for a stone to strike the ground if released from a height of 385 feet.

48. If an amount, a, is doubled n times, its final value is $a \cdot 2^n$.   A person wagers 5 cents on June 1 and doubles his wager each day for the balance of the month.   What is the amount of the wager on the last day of the month?

49. The area of an equilateral triangle of side s is given by $A = \dfrac{s^2 \sqrt{3}}{4}$.
Find the area of the equilateral triangle whose side is 5.19 inches long.

50 Interpolation in Logarithmic Computations

Since the mantissas in Table II are approximated to four figures we can, by interpolation (Section 35), determine the logarithm of a number containing four significant figures.   It will be recalled that tabular interpolation is based on the assumption that the rate of change of the function is constant throughout the interval under consideration. Although this is not strictly true in the case of logarithms, the results are accurate enough to be useful.

EXAMPLE 1.   Find the logarithm of 3.142 by interpolation.   Let m represent the mantissa of the number sequence 3142.   Using Table II we have

$$\text{Interval } A$$

Digits in number N	3140	3142	3150
Mantissa of log N	4969	m	4983

$$\text{Interval } B$$

We assume that the average rate of change in interval A equals the average rate of change in interval B.   Therefore

$$\frac{14}{10} = \frac{m - 4969}{2}$$

$$m - 4969 = \frac{2}{10} \cdot 14 = 3$$

$$m = 4969 + 3 = 4972$$

The last two equations indicate a procedure for making the interpolation mentally.   The mantissa m is seen to be a number which is $\frac{2}{10}$ of the way from 4969 to 4983.   Thus we add $\frac{2}{10}$ of the tabular difference 14 to the smaller mantissa to obtain the required mantissa.

In view of the element of approximation involved, Table II cannot be used to determine the logarithm of a number having more than four significant figures.   Such numbers should first be rounded off to four figures.   Thus to obtain the logarithm of 83.268 by the use of Table II, it is recommended that we merely evaluate the logarithm of 83.27.

Interpolation is also essential in determining an antilogarithm to four significant figures when the exact mantissa is not listed in the table. The procedure is shown in the following example.

EXAMPLE 2.   Determine the number whose logarithm is 0.8045 − 2. Let x represent the four digit sequence in the required antilogarithm. Using Table II, we find the following correspondence of values:

Interval A

Digits in number N	6370	x	6380
Mantissa of log N	8041	8045	8048

Interval B

Assuming the rate of change to be equal in intervals A and B, we have

$$\frac{7}{10} = \frac{4}{x - 6370}$$
$$x - 6370 = \tfrac{4}{7} \cdot 10 = 6$$
$$x = 6370 + 6 = 6376$$

Since the characteristic is −2, the required antilogarithm is 0.06376. Note that the number x is seen to be $\frac{4}{7}$ of the way from 6370 to 6380. Therefore the interpolation can be performed mentally by adding $\frac{4}{7}$ of the difference 10 to the smaller number.

Exercise 50

INTERPOLATION IN LOGARITHMIC COMPUTATION

Using Table II, find the logarithm of each of the following numbers:

1. 4081	**2.** 0.6237	**3.** 58,230
4. 0.03076	**5.** 279.9	**6.** 100,100
7. 8.268	**8.** 0.007155	**9.** 90.17

In each of the following, find the number N to four significant figures:

10. $\log N = 3.4208$

11. $\log N = 0.5744 - 3$

12. $\log N = 1.9612$

13. $\log N = 5.8661$

14. $\log N = 2.7705$

15. $\log N = 0.6384 - 1$

16. $\log N = 0.1442 - 4$

17. $\log N = 4.4848$

18. $\log N = 0.0055 - 2$

Evaluate each of the following expressions to four significant figures:

19. $(0.03764)^3$

20. $\dfrac{28.67}{1.983}$

21. $(748.5)(0.008722)$

22. $\sqrt[5]{3.261}$

23. $\dfrac{371.6\sqrt{409.3}}{84,830}$

24. $\dfrac{(0.1809)(3.265)}{\sqrt[4]{85.14}}$

25. $\dfrac{(2.148)(4.632)(0.3937)}{(24.35)(0.5756)}$

26. $\sqrt{\dfrac{(8.357)(6.342)}{359.6}}$

27. $\sqrt[3]{\dfrac{3.427}{(7.788)(2.129)}}$

28. $\dfrac{(16.31)^2(8.083)}{(0.0206)^2}$

29. $\dfrac{2.357}{(1.045)^{12}}$

30. $1000(1.015)^{10}$

31. The diagonal of a cube is $e\sqrt{3}$ where e is the length of an edge. Find the diagonal of a cube whose edge is 4.427 inches long.

32. The weight w in tons that will crush a certain type of structural column is given by the formula

$$w = 149.8\,\frac{d^{3.55}}{l^2}$$

where d is the diameter of the column in inches and l is its length in feet. What weight will crush a column 7 feet long and 4 inches in diameter?

33. For each 1000 bacteria in a culture, the number, n, to which they multiply after t hours is

$$n = 1000(10)^{1.637t}$$

Determine the value of n when t is 5.

34. A formula used to approximate the distance from which an object of given height is visible above the surface of the earth is

$$D = 1.320\,\sqrt{h}$$

where D is the distance in miles and h is the height in feet.   If the light of a lighthouse is 140.0 feet above sea level, for how many miles at sea is it visible?

35. The area of a sector of a circle whose central angle is x degrees is

$$A = \frac{\pi r^2 x}{360}$$

where r is the radius of the circle in any linear unit and A is the area expressed in the same unit squared. What is the area of a sector whose central angle is 18 degrees 15 minutes if the radius of the circle is 46.97 feet? (18 degrees 15 minutes = 18.25 degrees)

36. According to Kepler's third law, a planet will make one complete revolution about the sun in T years, where

$$T = a^{3/2}$$

and a is the average distance of the planet from the sun in astronomical units (1 A.U. = 93,000,000 miles approximately). How long does the planet Pluto require for one complete revolution if it is 39.52 A.U. from the sun?

37. The electrical resistance of a copper wire at 20°C. is given by the formula

$$R = \frac{10.37l}{d^2}$$

where R is the resistance in ohms, l is the length in feet, and d is the diameter in mils (1 mil = $\frac{1}{1000}$ inch). Find the resistance of 1 mile of copper wire whose diameter is 0.2576 inch.

38. If a circle is inscribed in a triangle, its radius is

$$r = \sqrt{\frac{(s-a)(s-b)(s-c)}{s}}.$$

where a, b, and c are the lengths of the sides and s, the semiperimeter, is $\frac{a+b+c}{2}$. If the sides of a triangle measure 28.61 inches, 24.31 inches, and 20.16 inches, what is the radius of its inscribed circle?

CHAPTER ■ 10

Variation

51 Ratio and Proportion

The Ratio of Two Numbers

In many practical problems it is necessary to compare the magnitude of two quantities. We may, for example, compare the speed of two cars traveling at 60 miles per hour and 30 miles per hour, respectively, by saying that the first car is traveling 30 miles per hour faster than the second car or by saying that the first car is traveling at twice the speed of the second car. The first comparison states the *difference* of the rates, that is, $60 - 30 = 30$. The second comparison states the *ratio* of the rates, that is, $60/30 = 2$.

In general, we define the ratio of one number a to another number b to be the quotient a/b. This ratio may be an abstract number or a rate, depending on the quantities involved. Thus 60 miles per hour/30 miles per hour = 2. On the other hand, the ratio of 165 miles to 3 hours, expressed as $\dfrac{165 \text{ miles}}{3 \text{ hours}}$, is the rate 55 miles per hour.

EXAMPLE 1. Find the ratio of 3 pounds to 5 ounces. Since 3 pounds equal 48 ounces, we have 48/5 as the ratio. This ratio may be read as "48 to 5."

The Meaning of Proportion

Suppose that a 3-inch by 5-inch photograph is enlarged. The optical system of the enlarger operates to give a faithful reproduction, on a larger

244

scale, of the original; the various components of the enlargement are in the same proportion as in the original.   If the width of the enlargement is 6 inches, what will the length be?   From our own experience, we know that the length will be 10 inches.   Note that the ratio of the widths of the two pictures is 3/6, and the ratio of the length is 5/10; in each case, the ratio equals 1/2.   The mathematical definition of proportion states these ideas in precise language; a proportion is a statement expressing the equality of two ratios.   Symbolically, the relation

$$\frac{a}{b} = \frac{c}{d} \tag{1}$$

is a proportion.   Referring again to the enlargement, the proportion may be written $w_1/w_2 = l_1/l_2$, where w_1 and l_1 are the width and length of the original photograph, and w_2 and l_2 is the width and length of the enlargement.

Terms of a Proportion

In the proportion (1), b and c are called the *means* of the proportion, and a and d are called the *extremes*.   The quantity d is called the *fourth proportional* to a, b, and c.   If the proportion is of the form

$$\frac{a}{x} = \frac{x}{d} \tag{2}$$

x is called the *mean proportional* between a and d, and d is called the *third proportional* to a and x.

EXAMPLE 2.   6 is the mean proportional between 4 and 9, since $4/6 = 6/9$.

A Fundamental Law of Proportion

Although the proportion (1) may be treated as a fractional equation, it occurs so frequently in applied mathematics that it is convenient to develop a more direct method of working with it.   If we multiply both members of (1) by bd, the result is

$$ad = bc \tag{3}$$

Expressed in words, Equation 3 states:

In a proportion, the product of the means equals the product of the extremes.

EXAMPLE 3.   Find the mean proportional between 2 and 32.

$$\frac{2}{x} = \frac{x}{32}$$
$$x^2 = 64$$
$$x = +8 \text{ or } -8$$

Problems Solved by Proportion

In many problems the functional relationship of the variables is most conveniently expressed as a proportion.   For example, Hooke's law states that the extension of a spring is proportional to the applied force. This functional relationship takes the form of the proportion

$$\frac{s_1}{s_2} = \frac{F_1}{F_2}$$

where s_1 is the extension due to a force of F_1 pounds, and s_2 is the extension due to a force of F_2 pounds.

EXAMPLE 4.   If a 10 pound weight causes a spring to stretch 3 inches, what weight will cause a stretch of 5 inches?   The corresponding proportion is

$$\frac{3}{5} = \frac{10}{F_2}$$
$$3F_2 = 50$$
$$F_2 = 16\tfrac{2}{3} \text{ pounds}$$

Therefore a weight of $16\tfrac{2}{3}$ pounds will cause a stretch of 5 inches in the spring.

Exercise 51

RATIO AND PROPORTION

Which of the following statements express comparison in the form of a ratio?

1. The temperature dropped 5° in the last hour.

2. The college enrollment is expected to double in the next 10 years.

3. During the month of February, a 20% reduction in prices will be given.

4. The postage of a letter increased 1 cent in 1958.

5. The budget is expected to be 20% higher this year.

6. The price of a 2-pound can of coffee is 10 cents lower during the sale.

7. The new school building will serve twice as many students as the old building.

8. The new edition of the book has two more chapters.

9. The new gun has twice the range of the older model.

10. Her typing speed increased by 10 words per minute.

11. His watch gains 2 minutes every day.

12. He doubled his income after finishing night school.

13. The tax rate is $1.02 per $100 assessed valuation.

14. A mixture of paint contains 1 part blue to 8 parts of white.

Find the ratio of the following quantities:

15. 2 gallons to 3 quarts	**16.** 2 hours to 45 minutes
17. 2 feet to 7 feet	**18.** 7 feet to 2 feet
19. 3 yards to 4 feet	**20.** 5 ounces to 3 pounds
21. 20 minutes to 3 hours	**22.** 4 inches to 3 feet
23. 2 yards to 3 inches	**24.** 20 cents to $1.50
25. 2 tons to 300 pounds	**26.** 1200 feet to 1 mile
27. 440 yards to 1 mile	**28.** 30 seconds to $2\frac{1}{2}$ minutes
29. 300° to 2 revolutions	**30.** 1 mile to 440 yards
31. 2 pecks to 7 bushels	**32.** 1 cubic foot to 1 cubic inch

Solve each one of the following proportions:

33. $\dfrac{2}{x} = \dfrac{3}{5}$ **34.** $\dfrac{x}{3} = \left(\dfrac{5}{4}\right)^4$

35. $\dfrac{17}{5} = \dfrac{3}{x}$ **36.** $\dfrac{1}{R} = \dfrac{3^2}{12^2}$

37. $\dfrac{2}{5} = \dfrac{x}{12}$ **38.** $\dfrac{7}{x} = \left(\dfrac{2}{5}\right)^2$

39. $\dfrac{9}{x} = \dfrac{x}{4}$ **40.** $\dfrac{9}{16} = \left(\dfrac{6}{t}\right)^2$

41. $\dfrac{3}{7} = \dfrac{x}{9}$ **42.** $\dfrac{64}{27} = \dfrac{125}{x^3}$

Find the mean proportional between each of the following pairs of numbers or letters.

43. 2, 128	**44.** 3, 27	**45.** 8, 2
46. 2, 72	**47.** 45, 3	**48.** 18, 2
49. a, b	**50.** x, y	**51.** a^2, b^2
52. a^3b, ab^3	**53.** $2a, 8a^3$	**54.** $2x, 32x$

Develop each of the following properties of proportions:

55. If $\dfrac{a}{b} = \dfrac{c}{d}$, then $\dfrac{a}{c} = \dfrac{b}{d}$. The property of mean alternation.

56. If $\dfrac{a}{b} = \dfrac{c}{d}$, then $\dfrac{b}{a} = \dfrac{d}{c}$. The property of inversion.

57. If $\dfrac{a}{b} = \dfrac{c}{d}$, then $\dfrac{a+b}{b} = \dfrac{c+d}{d}$. The property of addition.

Hint: Add 1 to each member of $\dfrac{a}{b} = \dfrac{c}{d}$.

58. If $\dfrac{a}{b} = \dfrac{c}{d}$, then $\dfrac{a-b}{b} = \dfrac{c-d}{d}$. The property of subtraction.

59. Using the proportion $\frac{2}{3} = \frac{8}{12}$, verify each of the properties 55 to 58.

Solve the following problems by proportion.

60. A tree 30 feet tall casts a shadow of 8 feet. At the same time, a flagpole casts a shadow 6 feet. Find the height of the flagpole.

61. An airplane uses 150 gallons of gasoline in a flight of 200 miles. At the same rate, how many gallons would be consumed in a flight of 750 miles?

62. A picture 3 inches by 5 inches is enlarged so that it is 12 inches long. What is the width?

63. About 2.1 pounds of butterfat can be obtained from 100 pounds of whole milk. How much butterfat is contained in a quart (32 ounces) of milk?

64. An insect spray contains 3 pounds of arsenate of lead dissolved in 50 gallons of water. How much arsenate of lead should be dissolved in a gallon of water?

65. If a certain solder consists of 2 parts of tin to 3 parts of lead, what amounts of each are necessary to make 25 pounds of solder?

52 Direct Variation

The Direct Variation $y = kx$

The functional relationship expressed by a proportion is often more conveniently expressed in a symbolic form called *variation*. We define direct variation as follows:

If the functional relationship of two variables, x and y, is such that the ratio of y to x for all pairs of values is constant, y varies directly as x. Symbolically, direct variation is the functional relationship $y/x = k$, or, as it is usually written,

$$y = kx \tag{1}$$

The constant k is called the *constant of variation*, and it may be determined mathematically or experimentally. In general, the direct variation $y = kx$ expresses the fact that as x increases, y increases. We may also express the relation (1) as y is directly proportional to x or y varies directly as x.

EXAMPLE 1.   It was known even in ancient times that the circumference of a circle varied directly as the diameter, that is, $C = kd$. Frequently k was taken as 3. As we know, the universal symbol for k in this case is π, so that we write $C = \pi d$. Although the exact value of π can never be measured, it is possible to calculate its value mathematically to any desired degree of accuracy. For many purposes, 3.1416 is sufficiently accurate.

EXAMPLE 2.   If a car travels at a constant rate, the distance varies directly as the time. The equation is therefore $d = kt$. In this case. the constant of variation represents the rate of travel, so that the equation is usually written $d = rt$.

Direct Variation as Proportion

Suppose that the direct variation $y = kx$ is satisfied by the pair of values (x_1, y_1). Then we have

$$y_1 = kx_1 \tag{2}$$

If (x_2, y_2) is another pair of values satisfying this relation, then

$$y_2 = kx_2 \tag{3}$$

If we now divide Equation 1 by Equation 2, the result is the proportion

$$\frac{y_1}{y_2} = \frac{x_1}{x_2} \tag{4}$$

It is clear, therefore, that the two relations

$$y = kx \tag{1}$$

$$\frac{y_1}{y_2} = \frac{x_1}{x_2} \tag{4}$$

are equivalent expressions for the same functional relationship between x and y. The following examples will show the particular advantage of each expression.

EXAMPLE 3. It is known that y varies directly as x, and that y equals 33 when x equals 12. What is the value of y when x equals 16?

Since $y = kx$, we have from the given data

$$33 = (k)(12)$$

Solving for k

$$k = \frac{33}{12} = \frac{11}{4}$$

Therefore

$$y = \frac{11}{4} x$$

When $x = 16$

$$y = \frac{11}{4} (16) = 44$$

Note that by using Equation 1, we find the specific relation between y and x, that is, $y = \frac{11}{4}x$, and we may then find any other pair of values from it.

EXAMPLE 4. Solve Example 3 by using Equation 4. Since $y_1/y_2 = x_1/x_2$, we have, using the given data,

$$\frac{33}{y_2} = \frac{12}{16}$$

Solving for y_2

$$y_2 = \frac{(16)(33)}{12} = 44$$

Note that although the solution is more direct, we do not find the specific functional relationship; we simply find the particular value of y when x equals 16.

The Direct Variation $y = kx^2$

When two variables, x and y, are related by the equation $y = kx^2$, we say that y varies directly as the square of x, or y is directly proportional to the square of x. This type of relationship is found frequently in geometric formulas and applied mathematics. Experiments by Galileo confirmed his belief that the distance s a freely falling body falls varies directly as the square of the time t, that is, $s = kt^2$. The value of k is usually denoted by $\frac{1}{2}g$, so that we have the formula

$$s = \tfrac{1}{2}gt^2$$

Experiments have shown that if the distance is measured in feet, and the time in seconds, then g, the acceleration due to gravity, is approximately equal to 32.2

It should be noted that the two equations

$$y = kx^2 \tag{5}$$

$$\frac{y_1}{y_2} = \left(\frac{x_1}{x_2}\right)^2 \tag{6}$$

are equivalent expressions for the same functional relationship between x and y.

EXAMPLE 5.   The areas of two similar polygons vary directly as the squares of any two corresponding sides.   If two similar polygons have two corresponding sides 10 inches and 4 inches, respectively, find the area of the second polygon if the area of the first is 250 square inches. Using Equation 5 written as $A = ks^2$, we have

$$250 = k(10)^2$$

or

$$k = 2.5$$

Then

$$A = 2.5s^2$$

When $s = 4$,

$$A = 2.5(4)^2 = 40$$

Therefore the area of the second polygon is 40 square inches.   We may, of course, use Equation 6 to get the same result.

Other Types of Direct Variation

Physics, chemistry, and other fields of applied mathematics furnish many other types of direct variation.   One important type has the symbolic form $y = k\sqrt{x}$, that is, y varies directly as the square root of x or y is directly proportional to the square root of x.   As an example, we have the well known law in physics: the time of oscillation of a pendulum varies directly as the square root of its length.   Symbolically, we have

$$T = k \sqrt{l} \tag{7}$$

$$\frac{T_1}{T_2} = \sqrt{\frac{l_1}{l_2}} \tag{8}$$

EXAMPLE 6.   A pendulum 36 inches long makes one complete oscillation in 1.92 seconds.   Find the length of a pendulum which makes one oscillation in 0.96 second.   Using Equation 8, we have

$$\frac{1.92}{0.96} = \sqrt{\frac{36}{l_2}} = \frac{\sqrt{36}}{\sqrt{l_2}}$$

$$2 = \frac{6}{\sqrt{l_2}}$$

$$\sqrt{l_2} = 3$$

$$l_2 = 9$$

The second pendulum is 9 inches long.

Exercise 52

DIRECT VARIATION

Write each of the following examples of direct variation as an equation, using k for the constant of variation.   Write, also, the equivalent proportion for each example.

1. The circumference C of a circle varies directly as the radius r. What is the value of k?

2. The interest i on a certain unpaid bill varies directly as the time t.   What is the meaning of k?

3. The weight w of a given size of pipe varies directly as the length l. What is the meaning of k?

4. The perimeters p of two similar polygons are directly proportional to the lengths s of any two corresponding sides.

5. The height h of mercury in a thermometer varies directly as the temperature t.

6. The cost c of a turkey is directly proportional to the weight w. What is the meaning of k?

7. The area A of an equilateral triangle varies directly as the square of a side s.   What is the value of k?

8. The wind resistance R to the motion of a projectile varies directly as the square of the speed.

9. The area A illuminated on the screen by a movie projector varies as the square of the distance d from the screen.

10. The pitch p of a violin string varies directly as the square root of the tension t.

11. The volume V of a sphere varies directly as the cube of the radius r. What is the value of k?

12. The force F of the wind on a sail varies directly as the square of the velocity v of the wind.

13. The power P required to propel a ship is directly proportional to the cube of the velocity v.

14. The pressure P on a submarine is directly proportional to the depth d of the submarine below the surface.

15. The safe load L for a circular pillar is directly proportional to the fourth power of its diameter d.

16. The power P in an electrical circuit of given resistance varies directly as the square of the current I.

17. Kepler's third law of planetary motion states that the square of the time t required for a planet to complete its orbit around the sun varies directly as the cube of the distance d from the sun.

18. The current I in an electrical circuit varies directly as the voltage V.

Find the value of k in each of the following examples of direct variation. Give the specific equation, and find the value of the indicated variable.

19. y varies directly as x, and $y = 8$ when $x = 2$. Find y when $x = 4$.

20. y varies directly as x, and $y = \frac{1}{2}$ when $x = 10$. Find x when $y = 8$.

21. y varies directly as the square of x, and $y = 4$ when $x = 2$. Find y when $x = 4$.

22. y varies directly as the square of x, and $y = \frac{2}{3}$ when $x = \frac{1}{2}$. Find x when $y = \frac{1}{2}$.

23. y varies directly as the square root of x, and $y = 4$ when $x = 8$. Find x when $y = 4\sqrt{2}$.

24. y varies directly as the square root of x, and $y = 6$ when $x = 8$. Find y when $x = 16$.

25. y varies directly as the $\frac{3}{2}$ power of x, and $y = 120$ when $x = 16$. Find y when $x = 4$.

Use the proportion form of direct variation in each of the following problems:

26. y varies directly as x and $y = 8$ when $x = 2$. Find (*a*) y when $x = 5$ and (*b*) x when $y = 192$.

27. y varies directly as x and $y = \frac{2}{3}$ when $x = 4$. Find (*a*) y when $x = 1$ and (*b*) x when $y = \frac{1}{3}$.

28. y varies directly as the square of x and $y = 16$ when $x = 2$. Find (*a*) y when $x = 6$ and (*b*) x when $y = 4$.

29. y varies directly as the square root of x and $y = 42$ when $x = 16$. Find (*a*) y when $x = 36$ and (*b*) x when $y = 21$.

30. In the direct variation $y = kx$, let $x = 4$ and $k = 10$.

(*a*) How is y changed when x is doubled?

(*b*) How is y changed when x is tripled?

(*c*) How is y changed when x is made half as large?

(*d*) How is y changed when x is made one-fourth as large?

31. In the direct variation $y = kx^2$, let $x = 4$ and $k = 10$. How is y changed as in *a*, *b*, *c*, and *d* of Problem 30?

32. In the direct variation $y = k\sqrt{x}$, let $x = 4$ and $k = 10$.

(*a*) How is y changed when x is taken four times as large? Nine times as large?

(*b*) How is y changed when x is taken one-fourth as large? One-ninth as large?

33. Do the answers to Problems 30, 31, and 32 depend on the value of k? Do they depend on the initial value of x?

34. The pressure P on the bottom of a water tank varies directly as the height h of the water. If the pressure caused by 5 feet of water is 315 pounds per square foot:

(*a*) What is the specific equation?

(*b*) What is the pressure caused by 12 feet of water?

(*c*) What is the meaning of k?

35. The resistance R of a given size of wire at constant temperature varies directly as the length l. The resistance of 100 feet of number 14 copper wire is 0.253 ohm.

(*a*) What is the specific equation?

(*b*) Find the resistance of 20 feet of number 14 copper wire.

(*c*) Find the length of number 14 copper wire which will have a resistance of 1 ohm.

36. A falling body strikes the ground with a speed v which varies directly as the square root of the distance s it fell. A ball falling 100 feet strikes the ground with a velocity of 80 feet per second.

(*a*) What is the specific equation?

(*b*) What is the speed of a ball dropped from the Washington monument? (Approximately 555 feet.)

(*c*) Find the distance a ball fell if it struck the ground with a speed of 40 feet per second.

37. The pressure P of wind against a billboard varies directly as the square of the velocity v of the wind. When the velocity of the wind is 30 miles per hour, the pressure is 2 pounds per square foot.

(*a*) What is the specific equation?

(*b*) Find the pressure when the velocity is 45 miles per hour.

(*c*) Find the velocity necessary to cause a pressure of 8 pounds per square foot.

38. The distance s a body falls from rest varies directly as the square of the time t. The body falls 64 feet in 2 seconds.

(*a*) Find the distance it falls in 8 seconds.

(*b*) Find the time to fall 400 feet.

39. The distance d required to stop an automobile varies as the square of the speed v when the brakes are first applied. A certain car can stop in 70 feet from a speed of 40 miles per hour. What is the distance it takes to come to a stop if going 70 miles per hour?

40. If the reaction time of the driver of the car in Problem 39 is $\frac{3}{4}$ second, can he avoid hitting a stalled car 200 feet ahead if he is going 60 miles per hour?

41. The power to drive a ship varies directly as the cube of the speed. If a speed of 12 knots requires 5000 horsepower, what horsepower is needed for a speed of 20 knots?

42. Using the information in Example 6, what is the length of a pendulum which makes one complete oscillation in 2 seconds?

53 Inverse Variation

The Inverse Variation $y = k/x$

A characteristic of the direct variation $y = kx$ is that as x increases, y increases proportionally. There are many functional relationships, however, in which an increase in one quantity causes a proportional decrease in another quantity. As an example, let us consider a rectangle with a given area. If we increase the length, we shall have to decrease the width if the area is to remain the same. Let l_1 and w_1 be the dimensions of the first rectangle and l_2 and w_2 the dimensions of the second rectangle. Since the area of a rectangle equals the product of the length by the width,

$$A = l_1 w_1$$
$$A = l_2 w_2$$

Since the areas are equal, we have

$$l_1 w_1 = l_2 w_2$$

$$\text{or} \qquad \frac{w_1}{w_2} = \frac{l_2}{l_1} \tag{1}$$

Note how Equation 1 differs from the relationship of two similar rectangles, where

$$\frac{w_1}{w_2} = \frac{l_1}{l_2}$$

When two variables, x and y, are related as in Equation (1) so that

$$\frac{y_1}{y_2} = \frac{x_2}{x_1} \tag{2}$$

we say that y *varies inversely as* x. This relation is equivalent to the relation

$$y = \frac{k}{x} \tag{3}$$

which is the usual definition of inverse variation. In Problem 1 of Exercise 53, the student will be asked to show that Equations 2 and 3 are equivalent.

EXAMPLE 1. Boyle's Law in chemistry states that the volume of a gas at constant temperature varies inversely as the pressure applied. If the volume of a gas is 60 cubic inches when the pressure is 80 pounds, what will the volume be when the pressure is 40 pounds? Since $V = k/p$, we have from the given data $60 = k/80$. Solving for k, $k = 4800$; therefore $V = 4800/p$, the specific relationship. When $p = 40$, $V = 4800/40 = 120$ cubic inches.

We may also use the equivalent relation

$$\frac{V_1}{V_2} = \frac{P_2}{P_1}$$

Substituting the given data

$$\frac{60}{V_2} = \frac{40}{80}$$

Solving for V_2

$$V_2 = \frac{(80)(60)}{40} = 120 \text{ cubic inches, as before.}$$

Other Types of Inverse Variation

The relation $y = k/x^2$, read "y varies inversely as the square of x," and $y = k/\sqrt{x}$, read "y varies inversely as the square root of x," are among the forms of inverse variation found in applied mathematics. An important example of the first type is the force of attraction between two given bodies, which varies inversely as the square of the distance between them. Thus

$$F = \frac{k}{d^2}$$

From this basic relation, it is possible to calculate by means of advanced mathematics the so called "velocity of escape," which is the velocity that must be imparted to a body so that it does not return to the earth.

It should be noted that the two equations

$$y = \frac{k}{x^2} \tag{4}$$

$$\frac{y_1}{y_2} = \left(\frac{x_2}{x_1}\right)^2 \tag{5}$$

are equivalent expressions for inverse variation and that

$$y = \frac{k}{\sqrt{x}} \tag{6}$$

$$\frac{y_1}{y_2} = \sqrt{\frac{x_2}{x_1}} \tag{7}$$

are also equivalent.

EXAMPLE 2.   If y varies inversely as the square of x, and $y = 81$ when $x = 3$, find y when $x = 9$.   Since $y = k/x^2$, we have from the given data $81 = k/9$.   Solving for k, $k = 9(81) = 729$.   Therefore $y = 729/x^2$, the specific relationship.   When $x = 9$, $y = 729/81 = 9$.

We may also use the equivalent relation

$$\frac{y_1}{y_2} = \left(\frac{x_2}{x_1}\right)^2$$

Substituting the given data

$$\frac{81}{9} = \left(\frac{x^2}{3}\right)^2$$

Solving for x_2

$$(x_2)^2 = \frac{(81)(9)}{9}$$

or $x_2 = 9$, as before.

Exercise 53

INVERSE VARIATION

1. Show that Equations 2 and 3 are equivalent expressions of inverse variation.

2. Show that Equations 4 and 5 are equivalent; show also that Equations 6 and 7 are equivalent.

Write each of the following examples of inverse variation as an equation, using k for the constant of variation.   Write also the equivalent proportion for each example.

3. The weight w of an object above the surface of the earth is inversely proportional to the square of the distance s from the center of the earth.

4. The electrical resistance R of a copper wire of given length varies inversely as the cross sectional area A.

5. The intensity of illumination I from a projector varies inversely as the square of the distance d to the screen.

6. The current I in an electrical circuit with given voltage varies inversely as the resistance R.

7. The acceleration of gravity g of a body varies inversely as the square of its distance s from the center of the earth.

8. The force F of attraction between two electrically charged particles varies inversely as the square of the distance s between them.

9. The time T to travel a given distance varies inversely as the average speed V.

10. The base b of a triangle of given area varies inversely as the altitude h. What is the value of k?

11. The height h of a cylinder of given volume varies inversely as the square of the radius r. What is the value of k?

12. The centripetal force F of a body moving at constant speed in a circular path varies inversely as the radius r of the path.

13. The mechanical advantage A of an inclined plane of a given length varies inversely as the height h of the raised end.

14. On a balanced lever, the weight w is inversely proportional to the distance d from the fulcrum.

Find the value of k in each of the following examples of inverse variation. Give the specific equation, and find the value of the indicated variable.

15. y varies inversely as x, and $y = 8$ when $x = 2$. Find y when $x = 4$.

16. y varies inversely as x, and $y = \frac{1}{3}$ when $x = 8$. Find y when $x = 4$.

17. y varies inversely as the square of x, and $y = 6$ when $x = 3$. Find y when $x = 2$.

18. y varies inversely as the square of x, and $y = \frac{2}{3}$ when $x = \frac{1}{2}$. Find x when $y = 4$.

19. y varies inversely as the square root of x, and $y = 5$ when $x = 9$. Find y when $x = 36$.

20. y varies inversely as the square root of x, and $y = 16$ when $x = 144$. Find x when $y = 64$.

Use the proportion form of inverse variation in each of the following problems:

21. y varies inversely as x, and $y = 6$ when $x = 15$. Find (*a*) y when $x = 3$ and (*b*) x when $y = 18$.

22. y varies inversely as the square of x, and $y = 8$ when $x = 10$. Find (*a*) y when $x = 20$ and (*b*) x when $y = 32$.

23. y varies inversely as the square root of x, and $y = 25$ when $x = 100$. Find (*a*) y when $x = 16$ and (*b*) x when $y = 10$.

24. In the inverse variation $y = k/x$, let $k = 10$ and $x = 2$.

(*a*) How is y changed when x is doubled?

(*b*) How is y changed when x is tripled?

(*c*) How is y changed when x is taken one-half as large?

(*d*) How is y changed when x is taken one-fourth as large?

25. Do the answers to Problem 24 depend on the value of k? Do they depend on the initial value of x?

26. For the inverse variation $y = k/x^2$, with $k = 10$ and $x = 2$, answer questions (a), (b), (c), and (d) of Problem 24.

27. For the inverse variation $y = k/\sqrt{x}$, with $k = 10$ and $x = 2$, answer questions (a), (b), (c), and (d) of Problem 24.

28. Do the answers to Problems 26 and 27 depend on the value of k? Do they depend on the initial value of x?

29. The weight of a body above the surface of the earth varies inversely as the square of its distance from the center of the earth. A boy weighs 150 pounds on earth's surface. How much would he weigh 300 miles above the earth's surface? (Assume the radius of the earth to be 4000 miles.)

30. The intensity of illumination from a spotlight varies inversely as the square of the distance from the light. An object 40 feet from the light has an intensity of 100 foot-candles. What is the illumination when the object is 20 feet from the light?

31. A projector is 20 feet from a screen. How far from the screen should the projector be moved in order to double the illumination?

32. Assuming the acceleration of gravity of a body on earth is 32.2 feet per second squared, find the acceleration of gravity of a satellite 300 miles from the surface of the earth. (See Problem 7. Assume the radius of the earth to be 4000 miles.)

33. If a weight tied to a string is given a circular motion in a horizontal plane, the height from the point of suspension is inversely proportional to the square of the number of revolutions per second. If the distance is 9.78 inches when the weight is revolving at one revolution per second, what is the height when the weight revolves at $1\frac{2}{3}$ revolutions per second?

34. A cylindrical oil can has a diameter of 7.750 inches and a height of 5.375 inches. If it is desired to change the radius to 3 inches, what will the height be if the new can is to hold the same amount of oil? (See Problem 11.)

54 Combined Variation

In our study of variation thus far, we have considered only functional relationships between two variables. But many functional relationships of geometry, science, in fact of everyday life, involve more than two quantities.

EXAMPLE 1.   The distance traveled depends on both the average rate and the time.

EXAMPLE 2.   The interest earned on money in the bank depends on the principal, the rate of interest, and the time.

EXAMPLE 3.   The safe load on a horizontal beam depends on its width and thickness, and the distance between its supports.

Joint Variation

If the quantity z varies directly as the product of x and y, then we say that z varies jointly as x and y.   Symbolically

$$z = kxy \tag{1}$$

The proportion from of Equation 1 is

$$\frac{z_1}{z_2} = \frac{x_1 y_1}{x_2 y_2} \tag{2}$$

EXAMPLE 4.   The lateral area of a right circular cylinder varies jointly as the radius and the height, that is $L = krh$.   In this example, we know that k equals 2π, so that we have the familiar formula $L = 2\pi rh$.

EXAMPLE 5.   The volume of a right circular cylinder varies jointly as the square of the radius and the height, that is, $V = kr^2h$.   In this example, k equals π, so that we have another familiar formula, $V = \pi r^2 h$.

Combined Variation

In Example 3, it seems reasonable that increasing either the width or the thickness of a beam will increase the safe load, and increasing the distance between supports will decrease the safe load.   This situation represents *combined variation:* the safe load of a horizontal beam varies jointly as the width and the square of the thickness, and inversely as the distance between supports.   Symbolically

$$L = \frac{kwh^2}{d}$$

Thus the functional relationships

$$z = \frac{kx}{y} \tag{3}$$

$$\frac{z_1}{z_2} = \frac{x_1 y_2}{x_2 y_1} \tag{4}$$

represent combined variation.

EXAMPLE 6.   The centripetal force F acting on a car going around a circular curve varies jointly as the weight w of the car and the square of the velocity v, and inversely as the radius r of the curve.   Symbolically,

$$F = \frac{kwv^2}{r}$$

In this example, k equals $1/g$, so that we have

$$F = \frac{wv^2}{gr}$$

the formula for centripetal force acting on a body.

Exercise 54

COMBINED VARIATION

Write each of the following examples of joint or combined variation as an equation, using k for the constant of variation.   Write also the equivalent proportion for each example.

1. The pressure p of gas in a container varies jointly as the density d and the absolute temperature t.

2. The total force of the wind F on a wall varies jointly as the area A of the wall and the square of the velocity v of the wind.

3. The resistance R of a copper wire varies directly as the length l and inversely as the square of the diameter d.

4. The volume V of a right circular cone varies jointly as the altitude h and the square of the radius r of the base.

5. The lift L on an airplane wing varies jointly as the area A of the wing and the square of the velocity v of the plane.

6. The distance d traveled by a car varies jointly as the average velocity v and the time t.

7. What is the value of k in Problem 6?

8. The cost c of gasoline on a certain trip varies directly as the distance d in miles and inversely as the number of miles per gallon n the car consumes.

9. What does the constant in Problem 8 represent?

10. The number of vibrations v of a violin string varies directly as the square root of the tension t and inversely as the product of the length l and diameter d.

11. The illumination E in foot-candles of a screen varies directly as the intensity I of the source in candle power and inversely as the square of the distance d in feet from the source.

12. The force of friction F between the tires and road necessary to prevent a car from skidding when rounding a curve varies inversely as the radius r of the curve and directly as the square of the velocity v.

Find the value of k in each of the following examples of variation. Give the specific equation, and find the value of the indicated variable.

13. z varies jointly as x and y, and $z = 24$ when $x = 2$ and $y = 3$. Find z when $x = 3$ and $y = 5$.

14. z varies jointly as x and the square of y, and $z = 25$ when $x = 3$ and $y = 5$.   Find z when $x = 6$ and $y = 4$.

15. z varies directly as x and inversely as y, and $z = 2$ when $x = 4$ and $y = 14$.   Find z when $x = 3$ and $y = 7$.

16. z varies directly as x and inversely as the square of y, and $z = 1$ when $x = 10$ and $y = 5$.   Find z when $x = 4$ and $y = 3$.

17. z varies directly as x and inversely as the square root of y, and $z = 10$ when $x = 5$ and $y = 81$.   Find z when $x = 15$ and $y = 9$.

18. z varies jointly as x and y and inversely as w, and $z = 50$ when $x = 10$, $y = 5$, and $w = 2$.   Find z when $x = 5$, $y = 12$, and $w = 4$.

19. z varies jointly as x and y and inversely as the square of w, and $z = 40$ when $x = 2.5$, $y = 40$ and $w = 5$.   Find z when $x = 4.5$, $y = 60$ and $w = 10$.

20. z varies jointly as x and y and inversely as the square root of w, and $z = 600$ when $x = 2$, $y = 3$ and $w = 4$.   Find z when x and y both equal 6 and $w = 25$.

Use the proportion form of variation in each of the following problems:

21. z varies jointly as x and y and $z = 14$ when $x = 4$ and $y = 28$. Find (a) z when $x = 8$ and $y = 15$ and (b) x when $z = 2$ and $y = 4$.

22. z varies directly as x and inversely as y, and $z = 75$ when $x = 30$ and $y = 2$.   Find (a) z when $x = 20$ and $y = 10$ and (b) y when $x = 10$ and $z = 2$.

23. z varies directly as x and inversely as the square of y, and $z = 10$ when $x = 5$ and $y = 3$.   Find (a) z when $x = 32$ and $y = 4$ and (b) y when $z = 2$ and $x = 4$.

24. z varies directly as x and inversely as the square root of y, and $z = 10$ when $x = 5$ and $y = 36$. Find (a) z when $x = 3$ and $y = 9$ and (b) x when $y = 4$ and $z = 8$.

Solve each of the following problems:

25. The general gas law states that the volume of a gas varies directly as the absolute temperature and inversely as the absolute pressure. If 20 cubic feet of a gas at 300 degrees absolute temperature has a pressure of 760 millimeters, what volume will the gas occupy at 310 degrees absolute temperature and 800 millimeter pressure?

26. The force of the wind on a billboard varies jointly as the area of the surface and the square of the velocity of the wind. If the force on a billboard 6 feet × 12 feet is 50 pounds when the wind is blowing at 10 miles per hour, what is the force on a billboard 8 feet × 10 feet when the wind is blowing at 30 miles per hour?

27. The weight of a cast iron cylinder varies jointly as the height and the square of the radius. A cylinder 2 inches high with a radius of 1 inch weighs 1.64 pounds. What is the weight of a cylinder 8 inches high with a radius of 4 inches?

28. A beam 16 feet long, 3 inches wide, and 6 inches thick can safely hold a load of 1500 pounds. What is the safe load of a beam of the same material 12 feet long, 2 inches wide, and 4 inches thick?

29. If the 16 foot beam of problem 28 is to carry a 3000 pound load, what will the thickness be if the width remains at 3 inches?

30. If the 16 foot beam of problem 28 is to carry the same load of 1500 pounds, what will the thickness be if the width is reduced to $1\frac{1}{2}$ inches?

CHAPTER 11

Introduction to the Mathematics of Sets

55 Review and Extension of the Set Concept

Modern mathematics has introduced entirely new points of view which give some older parts of mathematics new meaning. Much of modern mathematics is founded on the concept of set. It is basic to contemporary mathematical thought in statistics, probability, mathematical logic, and the newer geometries. It has developed a way of thinking which is of great promise in many important fields of human interest. Although the application of set thinking to these fields is beyond the scope of this book, it will be our purpose to extend the concept of set as we have used it in various sections of this text and to show how it can be applied to an algebra of sets.

Solution Sets

From our study of quadratic equations, we know that the roots of the equation $x^2 + 2x - 8 = 0$ are $x = 2$ and $x = -4$. In the language of sets this can be expressed in the form

$$\{x \mid x^2 + 2x - 8 = 0\} = \{2, -4\}$$

which is read: "The set of all x's such that $x^2 + 2x - 8 = 0$ is the set $\{2, -4\}$."

In general, if $p(x)$ is a condition on x, then the set of all elements in U which satisfy the condition $p(x)$ is written

$$\{x \in U \mid p(x)\}$$

and is read: "The set of all x's in U such that $p(x)$ is true." The set of elements which satisfy the condition is called a *solution set*. Thus for the preceding example the elements 2 and -4 comprise a solution set. We will usually exhibit a solution set by listing its elements, although at times a graphic representation may be more appropriate or more desirable. Procedures for both of these methods of exhibiting a solution set are illustrated in the following examples.

EXAMPLE 1. If $A = \{1, 3, 5, 7, 9\}$, list the solution set $B = \{x \in A \mid x < 7\}$. The elements of set A which are less than 7 are 1, 3, 5. Therefore $B = \{1, 3, 5\}$.

EXAMPLE 2. If $R = \{$real numbers$\}$, graph the solution set $S = \{x \in R \mid x < 4\}$. If we represent an algebraic scale on a straight line, than all real numbers less than 4 correspond to all points on the line to the left of the point $+4$. Therefore we draw a half-line from the point $+4$ to the left. The open circle at the point $+4$ is used to indicate that this point is not included in the set of points. (See Figure 1.)

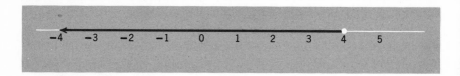

Figure 1

EXAMPLE 3. If $R = \{$real numbers$\}$, graph the solution set $T = \{x \in R \mid -1 < x \leq 3\}$. The set of all real numbers greater than -1 and less than or equal to $+3$ corresponds to the set of points on the line segment between the coordinates -1 and $+3$, excluding the point -1 as an end point. (See Figure 2.)

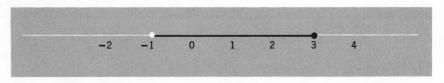

Figure 2

Universal Sets and Subsets

If A and B are sets which are so related that every object which belongs to A also belongs to B, then we say that A is contained in B, or that A is a *subset* of B. Thus the set $\{a, c, e\}$ is a subset of the set $\{a, b, c, d, e\}$, and the set of all rational numbers is a subset of the set of all real numbers. To indicate that A is a subset of B we write

$$A \subseteq B \qquad \text{or} \quad B \supseteq A$$

and to indicate that A is not a subset of B we write

$$A \nsubseteq B$$

The stated definition of subset implies that every set is a subset of itself. When it is important to specify that a subset of a set is not the set itself, we use the notation

$$A \subset B$$

which is read: "A is a *proper subset* of B."

EXAMPLE 4.   The set $\{3, 5\}$ is a proper subset of the set $\{1, 3, 5\}$.

EXAMPLE 5.   The set $P = \{3, 5, 1\}$ is not a proper subset of the set $Q = \{1, 3, 5\}$, although P is a subset of Q. We can write $P \subseteq Q$ or, more specifically, $P = Q$ since every element in P is an element in Q and conversely.

Another set which is considered to be a subset of every set is the so called *empty set*—the set which has no members. It is designated by the symbol ϕ. The extension of the set concept to cover the empty set is analagous to the extension of the system of directed numbers to include zero. Although it is not logically necessary to make this extension, it provides for greater generality and more convenient notation.

EXAMPLE 6.   The set of all rational numbers which are roots of the quadratic equation $x^2 - 5 = 0$ is an empty set.

The sets which we encounter in many situations are subsets of some very large set, such as the set of all real numbers or the set of all points in a plane. The large set is referred to as the *universal set* or *universe*. Of course the universal set may change from one context to another; it may be the set of real numbers, or the set of rational numbers, or the set of points on a line, etc. In each case, however, we shall denote the universal set by the symbol U. Thus

$$U = \{\text{universe}\}$$

Venn Diagrams

It is often helpful to visualize a set schematically by letting the points of a plane inside some closed curve represent the elements of the set. A *Venn diagram* is such a representation. In Figure 3 the rectangle U represents the universal set U. The elements of U correspond to points within the rectangle. A subset of U is denoted by a set of points enclosed by a circle (or any other convenient curve) within the rectangle, as for example the set of points within the circle A in the diagram. If A is a proper subset of B, this is indicated by enclosing the set of points A within the set of points B, as shown in Figure 3.

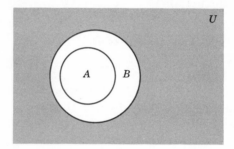

Figure 3

EXAMPLE 10. Construct a Venn diagram to illustrate the following subsets of the set of all people: all amateur athletes, all golfers, all professional male golfers. In Figure 4, U represents the set of all people, A the set of all amateur athletes, G the set of all golfers, and P the set

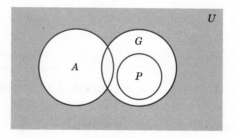

Figure 4

of all professional male golfers. P is a proper subset of G since all elements in P are elements in G, but not all elements in G belong in P. A and G are shown to overlap since there are some elements common to both A and G, namely, people who are amateur golfers.

Exercise 55

REVIEW AND EXTENSION OF THE SET CONCEPT

Let N represent the set of all positive integers less than or equal to 12. List the elements of the following sets.   Use braces.

1. Set N.
2. Subset T consisting of all multiples of 3 in N.
3. Subset B consisting of all elements in N greater than 7.
4. Subset A consisting of all solutions of $7x - 11 = 73$.
5. Subset C consisting of all solutions of $x^2 - 11x + 24 = 0$.

Write a verbal statement which means:

6. $K = \{x, y, z\}$
7. $P = \{1, 4, 9, 16, 25, 36, \ldots n^2\}$
8. $L = \{$Michigan, Huron, Ontario, Superior, Erie$\}$
9. $Q = \{-9, -7, -5, -3, -1\}$
10. $S = \{$Alaska, Hawaiian Islands$\}$
11. $G = \{$isosceles triangles, equilateral triangles, scalene triangles$\}$

Given $U = \{$real numbers$\}$.   Tell which of the following statements are true and which are false:

12. $5 \in U$
13. $-4 \in U$
14. $\pi \in U$
15. $\sqrt{3} \notin U$
16. $2.666 \ldots \notin U$
17. $\sqrt{-9} \notin U$
18. $\{0, 1, 2, 3, \ldots\} \subset U$
19. $\{13\} \subseteq U$
20. $\{\sqrt{4}, \sqrt{-4}\} \subset U$
21. $U \supseteq \{$rational numbers$\}$
22. $\phi \subset U$

23. How many subsets of the set $\{p, q, r\}$ have exactly two elements? List them.

24. List all the subsets of the set $\{3, 5, 7, 9\}$ which have exactly two elements.   How many are there?

25. List all possible subsets of the set $\{a, b, c\}$.   Include the set itself and the empty set ϕ.

What is the total number of subsets which can be formed from a set containing the following number of elements? Include the set itself and the empty set ϕ.

26. 2 elements   **27.** 3 elements   **28.** 4 elements   **29.** n elements

If $U = \{-3, -1, 0, 1, 3\}$, list the elements of each solution set. If there are no elements which satisfy the given statement, indicate that the subset is ϕ.

30. $\{x \in U \mid x \geq 0\}$ **31.** $\{x \in U \mid x \not> 0\}$
32. $\{x \in U \mid x < -3\}$ **33.** $\{x \in U \mid x \neq -3\}$
34. $\{x \in U \mid -1 \leq x < 3\}$ **35.** $\{x \in U \mid -3 \leq x \leq 3\}$
36. $\{x \in U \mid x > -3\}$ **37.** $\{x \in U \mid 1 < x < 3\}$

If $U = \{\text{real numbers}\}$, graph the set of points on a straight line which represent each of the following subsets:

38. $\{x \in U \mid x \geq 2\}$ **39.** $\{x \in U \mid 3x = 9\}$
40. $\{x \in U \mid x \neq 5\}$ **41.** $\{x \in U \mid -2 \leq x \leq 4\}$
42. $\{x \in U \mid x \not> 0\}$ **43.** $\{x \in U \mid x > 1\}$
44. $\{x \in U \mid 4x > 20\}$ **45.** $\{x \in U \mid x^2 + x - 12 = 0\}$
46. $\{x \in U \mid 3x + 2 \neq -4\}$ **47.** $\{x \in U \mid x^2 + x - 12 \neq 0\}$
48. $\{x \in U \mid x^2 + x - 12 > 0\}$ **49.** $\{x \in U \mid x^2 + x - 12 < 0\}$

If $U = \{\text{real numbers}\}$, $R = \{\text{rational numbers}\}$, and $I = \{\text{integers}\}$, identify the elements of each of the following solution sets:

50. $\{x \in I \mid 3x + 1 = 2x\}$ **51.** $\{x \in R \mid 3x + 1 = 2x\}$
52. $\{x \in U \mid 2 < x < 3\}$ **53.** $\{x \in I \mid 2 < x < 4\}$
54. $\{x \in I \mid 6x^2 - 7x + 2 = 0\}$ **55.** $\{x \in R \mid 6x^2 - 7x + 2 = 0\}$
56. $\{x \in U \mid 6x^2 - 7x + 2 = 0\}$ **57.** $\{x \in I \mid x < 0\}$

In the following, show by means of a Venn diagram the relations of the sets P, Q, and U if $U = \{1, 2, 3, 4, \ldots 25\}$ and:

58. $P = \{2, 4, 6, \ldots 24\}$, $Q = \{3, 6, 9, \ldots 24\}$
59. $P = \{x \in U \mid x < 5\}$, $Q = \{x \in U \mid x > 20\}$
60. $P = \{x \in U \mid x \neq 10\}$, $Q = \{x \in U \mid x \leq 6\}$
61. $P = \{1, 2, 3\}$, $Q = \{4, 5, 6, \ldots 25\}$
62. $P = \{x \in U \mid (x - 3)(x - 7) = 0\}$, $Q = \{3, 7\}$
63. $P = \{3, 5, 7, 11, 13, 17, 21, 23\}$, $Q = \{1, 3, 5, \ldots 25\}$

56 Cartesian Sets

When we list the roots of a quadratic equation, the order in which we present them is not important. Thus we can say that the roots of $x^2 - 5x + 6 = 0$ are 2 and 3 or 3 and 2. In many situations, however, the order in which two elements are presented is important. When the order is important we use a notation of the form $(—, —)$. The symbol (a, b) means that the two elements are to be presented in the order a first and b second. We call the elements a and b the coordinates of the *ordered pair* (a, b).

Now if A and B are sets, we can form another set whose members are all the ordered pairs whose first coordinates belong in A and whose second coordinates belong in B. For example, if $A = \{a, b\}$ and $B = \{2, 3, 4\}$, we can form the following ordered pairs

First element a	*First element b*
$(a, 2)$	$(b, 2)$
$(a, 3)$	$(b, 3)$
$(a, 4)$	$(b, 4)$

The set consisting of these 6 ordered pairs is called the *cartesian set of A and B* and is denoted by $A \times B$ (read: "A cross B").

EXAMPLE 1. Tabulate the cartesian set $A \times B$ if $A = \{a, b, c\}$ and $B = \{r, s, t, u\}$.

The ordered pairs can be extracted systematically by listing in separate columns the possible first elements and repeatedly pairing each first element with the possible selections of the second element. Thus

First element a	*First element b*	*First element c*
(a, r)	(b, r)	(c, r)
(a, s)	(b, s)	(c, s)
(a, t)	(b, t)	(c, t)
(a, u)	(b, u)	(c, u)

The cartesian set $A \times B$ consists of these 12 ordered pairs.

The Cartesian Set of U

As a special and very important case, suppose that sets A and B are both the same as a given set U. Then the cartesian set $A \times B$ is called the *cartesian set of U* and represents the set of all ordered pairs of elements in U.   It is denoted by $U \times U$.   If $U = \{3, 5, 7\}$, then

$U \times U$

$= \{(3, 3), (3, 5), (3, 7), (5, 3), (5, 5), (5, 7), (7, 3), (7, 5), (7, 7)\}$

Notice that all possible ordered pairs of the numbers 3, 5, 7 have been made to form the cartesian set.   The 3 elements in U lead to 3^2 or 9 elements in $U \times U$.

We can construct a graph of this set $U \times U$ in the usual way, using the first coordinate of the ordered pair as the abscissa of a point and the second coordinate as its ordinate.   The graph is shown in Figure 1.

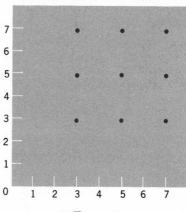

Figure 1

If U is the set of real numbers, then $U \times U$ is the set of all ordered pairs of real numbers.   Since these pairs may be placed in a one-to-one correspondence with the points in a plane, we may regard the plane as a geometric representation of the cartesian set of real numbers.   The development of analytic geometry rests on this assumption.

EXAMPLE 2.   If $U = \{1, 2, 3, 4\}$, list all the ordered pairs in $U \times U$ such that $y < x$.   If y is to be less than x, the only possible first elements are 2, 3, and 4.   With each of these we can form the following ordered pairs.

First element 2	*First element 3*	*First element 4*
(2, 1)	(3, 2)	(4, 3)
	(3, 1)	(4, 2)
		(4, 1)

Therefore the required pairs are the elements of the set

$$\{(2,\ 1),\ (3,\ 2),\ (3,\ 1),\ (4,\ 3),\ (4,\ 2),\ (4,\ 1)\}$$

Solution Sets of Ordered Pairs

If $p(x,\ y)$ is a condition on the variables x and y, the set of all ordered pairs which satisfy the condition is denoted by

$$\{(x,\ y) \in U \times U \mid p(x,\ y)\}$$

which is read: "the set of all ordered pairs in the cartesian set U cross U such that $p(x,\ y)$ is true." The solution set consists of all elements which satisfy the given condition.

EXAMPLE 3. Let $U = \{-3,\ -2,\ -1,\ 0,\ 1,\ 2,\ 3\}$. Find the solution set $\{(x,\ y) \in U \times U \mid y = x + 1\}$. Any element in U can be used as the first element of the ordered pair except $+3$. (For $x = +3$, the value of y is not an element in U.)

The required set is $\{(-3,\ -2),\ (-2,\ -1),\ (-1,\ 0),\ (1,\ 2),\ 2,\ 3)\}$.

Relations and Functions

In the mathematics of sets, a *relation* is defined to be a *set of ordered pairs*. A relation in $U \times U$ is therefore a subset of $U \times U$, and a relation in $A \times B$ is a subset of $A \times B$. We may denote a relation by the symbol $\textcircled{r}$, thus

$$\textcircled{r} = \{(x,\ y) \in A \times B \mid p(x,\ y)\}$$

The set of all the first coordinates of the ordered pairs belonging to $\textcircled{r}$ is called the domain of $\textcircled{r}$, and the set of all the second coordinates is called the range of $\textcircled{r}$. To illustrate, let $A = \{1,\ 2\}$ and $B = \{1,\ 2,\ 3\}$. Then the relation $\textcircled{r_1} = \{(x,\ y) \in A \times B \mid y > x\}$ consists of the set of ordered pairs $\{(1,\ 2),\ (1,\ 3),\ (2,\ 3)\}$. The domain of $\textcircled{r_1}$ is the set $\{1,\ 2\}$ and the range of r_1 is $\{2,\ 3\}$.

A *function* may now be defined as a special kind of relation. If the relation $\textcircled{r}$ is such that for every x in the domain of the relation there is exactly one corresponding y in the range of the relation, it is called a function. Thus the relation $\textcircled{r_1}$ is not a function since corresponding to the value $x = 1$ there are two different values of y. On the other hand the relation

$$\widehat{r_2} = \{(x, y) \in A \times B \mid y = x + 1\} = \{(1, 2), (2, 3)\}$$

is a function, since for each x in the domain there exists exactly one corresponding y in the range. When the relation is a function we use the notation

$$F(x)$$

to denote the second coordinate of the ordered pair whose first value is x. Thus y and $F(x)$ represent the same elements and we may write

$$y = F(x)$$

which is the notation employed in our previous study of functions. Thus we see that a function is merely one of a group of relations. This point will be more apparent and meaningful if we examine the graphic representation of certain relations.

Graphic Representation of Relations

The graph of the relation

$$\widehat{r} = \{(x, y) \in A \times B \mid p(x, y)\}$$

will be affected both by the composition of the sets A and B and the nature of the condition $p(x, y)$. The graph may consist of isolated points,

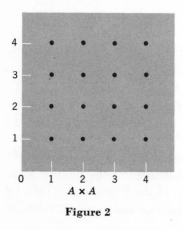

Figure 2

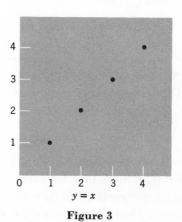

Figure 3

and if so we say that the range and domain are *discrete*. If, on the other hand, the graph of the relation is a smooth curve with no gaps, or an area with no gaps, the range and domain are said to be *continuous*. We now illustrate some of these possibilities.

Suppose $A = \{1, 2, 3, 4\}$. Then $A \times A$ is the set of points shown in Figure 2. Such an array of points, whose coordinates are integers, is known as a *lattice* of points. The relation

$$\textcircled{r_1} = \{(x, y) \in A \times A \mid y = x\}$$

consists of the ordered pairs $(1, 1)$, $(2, 2)$, $(3, 3)$, and $(4, 4)$. Its graph is shown in Figure 3. The relation

$$\textcircled{r_2} = \{(x, y) \in A \times A \mid y > x\}$$

consists of the ordered pairs $(1, 2)$, $(1, 3)$, $(1, 4)$, $(2, 3)$, $(2, 4)$, $(3, 4)$. The corresponding points are graphed in Figure 4. Figure 5 illustrates

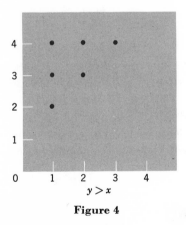

$y > x$

Figure 4

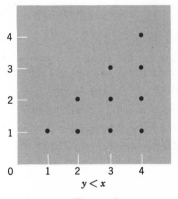

$y < x$

Figure 5

the graph of the relation

$$\textcircled{r_3} = \{(x, y) \in A \times A \mid y \leq x\}$$

It consists of the points $(1, 1)$, $(2, 2)$, $(2, 1)$, $(3, 3)$, $(3, 2)$, $(3, 1)$, $(4, 4)$, $(4, 3)$, $(4, 2)$, $(4, 1)$. In all four of these cases, the range and domain are discrete. Of the four, only the relation $\textcircled{r_1}$ (Figure 3) is a function.

Now consider the relation

$$\textcircled{r_4} = \{(x, y) \in U \times U \mid y = 2x\} \quad \text{where} \quad U = \{\text{real numbers}\}$$

This relation is represented graphically in Figure 6 by the set of points on the straight line $y = 2x$.

The relation r_5 = $\{(x, y) \in U \times U \mid y > 2x\}$ is represented in Figure 7. Note that the points in this set are indicated by an area above the line $y = 2x$, and that the shading lines stop short of $y = 2x$ to show that y must be greater than, and not equal to, $2x$.

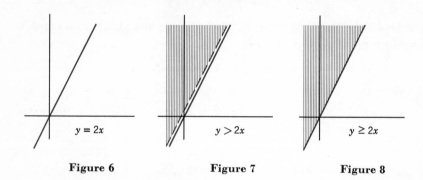

| Figure 6 | Figure 7 | Figure 8 |

The relation r_6 = $\{(x, y) \in U \times U \mid y \geq 2x\}$ is shown in Figure 8. In this example the shading lines intersect (but do not cross) the line $y = 2x$ to indicate that the set includes all points on the line $y = 2x$ as well as all points in the area above it. In each of these last three relations the range and domain are continuous. The relation r_4 is a function, since a single value of y corresponds to each value of x; the relations r_5 and r_6 are not functions since more than one value of y can be paired with a given value of x.

Exercise 56

CARTESIAN SETS

List all the elements of the cartesian set $A \times B$ if:

1. $A = \{a\}, B = \{r\}$
2. $A = \{1, 3\}, B = \{5\}$
3. $A = \{p, q\}, B = \{r, s\}$
4. $A = \{2, 3, 4\}, B = \{2, 5\}$
5. $A = \{0, 2, 4\}, B = \{1, 3, 5\}$
6. $A = \{a, b, c, d\}, B = \{r, s, t, u, v\}$

7. If the set A has m elements and the set B has n elements, how many ordered pairs does the set $A \times B$ have?

List all the elements of the cartesian set $A \times A$ if:

8. $A = \{a\}$ **9.** $A = \{u, v\}$
10. $A = \{-3, 0, 3\}$ **11.** $A = \{p, q, r, s\}$
12. $A = \{0, 2, 4, 6, 8\}$ **13.** $A = \{1, 2, 3, 4, 5, 6\}$
14. If set A has n elements, how many ordered pairs are in $A \times A$?

Let $U = \{1, 2, 3, \ldots 20\}$. Without listing the elements involved, determine each of the following:

15. How many elements are in the cartesian set $U \times U$?
16. How many elements in $U \times U$ have equal coordinates?
17. How many ordered pairs in $U \times U$ have a first coordinate whose value is 3?
18. How many elements in $U \times U$ have unequal coordinates?

If $U = \{1, 3, 5, 7\}$, list the elements of the cartesian set:

19. $\{(x, y) \in U \times U \mid y = x\}$ **20.** $\{(x, y) \in U \times U \mid y < x\}$
21. $\{(x, y) \in U \times U \mid y > x\}$ **22.** $\{(x, y) \in U \times U \mid y \neq x\}$
23. $\{(x, y) \in U \times U \mid y = x + 2\}$ **24.** $\{(x, y) \in U \times U \mid y = x + 1\}$

If $U = \{-3, -2, -1, 0, 1, 2, 3\}$, list the elements of the cartesian set:

25. $\{(x, y) \in U \times U \mid y \geq x\}$
26. $\{(x, y) \in U \times U \mid y = 2x + 1\}$
27. $\{(x, y) \in U \times U \mid y > 2x + 1\}$
28. $\{(x, y) \in U \times U \mid y - x = 3\}$

If $U = \{0, 1, 2, 3, 4, 5\}$, graph each of the following relations. State the range and domain of each relation.

29. $\{(x, y) \in U \times U \mid y = x\}$ **30.** $\{(x, y) \in U \times U \mid y \neq x\}$
31. $\{(x, y) \in U \times U \mid y \leq x\}$ **32.** $\{(x, y) \in U \times U \mid y > x\}$

If $U = \{-4, -3, -2, -1, 0, 1, 2, 3, 4\}$, graph each of the following subsets of $U \times U$:

33. $\{(x, y) \mid y = 2x + 1\}$ **34.** $\{(x, y) \mid y \leq 2x + 1\}$
35. $\{(x, y) \mid y > 2x + 1\}$ **36.** $\{(x, y) \mid y = -x\}$
37. $\{(x, y) \mid y = x^2\}$ **38.** $\{(x, y) \mid y > x^2\}$

If x and y are real numbers, graph each of the following relations. Indicate which are functions.

39. $\{(x, y) \mid y = x - 2\}$ **40.** $\{(x, y) \mid y \geq 2x + 2\}$
41. $\{(x, y) \mid y < \frac{1}{2}x + 1\}$ **42.** $\{(x, y) \mid y = x^2 - x - 2\}$
43. $\{(x, y) \mid y \leq 2x^2 - 2x\}$ **44.** $\{(x, y) \mid y - 1 > x + 4\}$
45. $\{(x, y) \mid y > 0 \text{ and } x < 0\}$ **46.** $\{(x, y) \mid y < 0 \text{ and } x < 0\}$

57   Operations with Sets

Just as numbers may be combined by certain operations (such as addition and multiplication) to form other numbers, sets may be combined by certain operations to form other sets. Three operations of particular importance to the study of sets are intersection, union, and complementation.

Intersection

The intersection of sets A and B, written $A \cap B$, is the set of elements which are members of both A and B. In symbolic notation

$$A \cap B = \{x \mid x \in A \text{ and } x \in B\}$$

which is read: "A intersection B is the set of all x's such that x is in A and x is in B." Three situations are possible.

1. A and B have no elements in common, as in Figure 1. The sets A and B are then said to be disjoint. Their intersection is an empty set, so in this example, $A \cap B = \phi$.

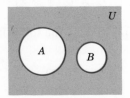

Figure 1

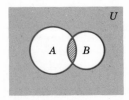

Figure 2

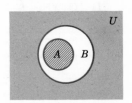

Figure 3

2. A and B have some members in common, as in Figure 2. $A \cap B$ is represented by the shaded portion of the diagram.

3. All members of one set are members of the other set.   The two sets may be equal or one may be a proper subset of the other.   Figure 3 illustrates the situation if A is a proper subset of B.   The shading indicates that, in this example, $A \cap B = A$.

EXAMPLE 1.   Represent each of the following statements by a Venn diagram:

(a) Tom Hawks is the only student at our college who is a member of both the chess club and the golf team.

(b) Each member of the golf team is also a member of the chess club, but some members of the chess club are not on the golf team.

(c) No member of the golf team belongs to the chess club.

If we let $U = \{$students at our college$\}$, $A = \{$members of golf team$\}$, $B = \{$members of chess club$\}$, statement (a) can be illustrated by the Venn diagram of Figure 2, statement (b) can be illustrated by Figure 3, and statement (c) by Figure 1.

EXAMPLE 2.   If $U = \{$real numbers$\}$, $A = \{x \in U \mid x \geq -1\}$, $B = \{x \in U \mid x < 7\}$, represent graphically the set $A \cap B$.   The graph of the set is shown in Figure 4.   It consists of all points common to the half-lines which represent sets A and B.

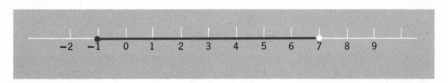

Figure 4

EXAMPLE 3.   If $U = \{$real numbers$\}$, $A = \{(x, y) \in U \times U \mid y = x^2\}$, $B = \{(x, y) \in U \times U \mid y = x + 6\}$, represent graphically the set $A \cap B$. In Figure 5, A is the set of all points on the parabola $y = x^2$. B is the set of all points on the line $y = x + 6$. $A \cap B$ is the set of points common to the parabola and line.   Specifically, $A \cap B = \{(-2, 4), (3, 9)\}$.

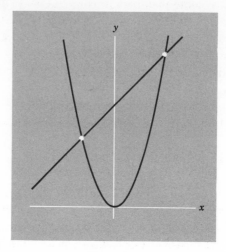

Figure 5

Union

The union of sets A and B, written $A \cup B$, is the set of elements which are members of either A or B (or both). We may write

$$A \cup B = \{x \mid x \in A \text{ or } x \in B\}$$

which is read: "A union B is the set of all x's such that x is in A or x is in B." The representation of $A \cup B$ by Venn diagrams for three possible situations is shown in Figures 6–8. In each example $A \cup B$ is shaded.

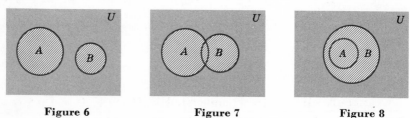

Figure 6   **Figure 7**   **Figure 8**

EXAMPLE 4. Represent graphically the set $A \cup B$ and the set $A \cap B$ in the universe of real numbers if $A = \{(x, y) \mid x \leq 0\}$ and $B = \{(x, y) \mid y \geq 2\}$. The subset A is represented in Figure 9 by the area to the left

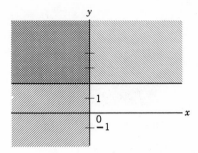

Figure 9

of the y-axis; the subset B is represented by the area above the line $y = 2$. $A \cup B$ is the entire shaded area since it contains all points in A, or B, or both. $A \cap B$ is the crosshatched area since it contains all the points which are in both A and B.

Complementation

The complement of set A is the set of all elements in the universal set U which are not in A.   We denote the complement of A by $\bar{A}$ which is read: "A bar."   Therefore

$$\bar{A} = \{x \in U \mid x \notin A\}$$

The set $\bar{A}$ is indicated by the shaded area in Figure 10.

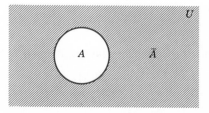

Figure 10

EXAMPLE 5.   Let $U = \{a, b, c, d, e, f, g\}$, $A = \{a, c, f\}$, $B = \{a, b, f, g\}$. The following listing illustrates how the operations of intersection, union, and complementation may be used to form new sets:

$$\bar{A} = \{b, d, e, g\} \qquad\qquad A \cap \bar{B} = \{c\}$$
$$\bar{B} = \{c, d, e\} \qquad\qquad \bar{A} \cup B = \{a, b, d, e, f, g\}$$
$$A \cap B = \{a, f\} \qquad\qquad \bar{A} \cap \bar{B} = \{d, e\}$$
$$A \cup B = \{a, b, c, f, g\} \qquad \overline{A \cap B} = \{b, c, d, e, g\}$$

Exercise 57

OPERATIONS WITH SETS

If $U = \{-3, -2, -1, 0, 1, 2, 3, 4\}$, $A = \{-3, -1, 1, 3\}$, $B = \{1, 2, 3, 4\}$, list the elements of the sets determined by the following operations:

1. $\bar{B}$ 2. $A \cup B$ 3. $A \cap B$
4. $\bar{A} \cup \bar{B}$ 5. $\bar{A} \cap B$ 6. $\overline{A \cup B}$

If $U = \{a, b, c, d, e, f, g, h\}$, $A = \{a, b, c, d, f\}$, $B = \{b, c, d\}$, list the elements of each of the following sets:

7. $A \cap B$ 8. $A \cup B$ 9. $\bar{A} \cap B$
10. $A \cap \bar{B}$ 11. $\overline{A \cap B}$ 12. $\bar{A} \cap \bar{B}$

13. Describe both the union and the intersection of the set of rational numbers and the set of irrational numbers.

14. Describe both the intersection and union of the set of integers and the set of rational numbers.

Use Venn diagrams to illustrate $A \cap B$ and $A \cup B$ if $A = \{a, b, c\}$ and:

15. $B = \{b, c, e\}$ 16. $B = \{b, c, a\}$
17. $B = \{e, f, g\}$ 18. $B = \{b\}$
19. $B = \{a, b, c, d, e\}$ 20. $B = \phi$

If $U = \{\text{real numbers}\}$, describe the complement of the set:

21. $\{x \in U \mid x > 3\}$ 22. $\{x \in U \mid x \geq -2\}$
23. $\{x \in U \mid x = 5\}$ 24. $\{x \in U \mid x \leq 0\}$

If $U = \{\text{real numbers}\}$, draw the graph of each of the following sets:

25. $A = \{x \in U \mid x > 3\}$ 26. $\bar{A}$
27. $B = \{x \in U \mid x \leq 4\}$ 28. $A \cap B$
29. $A \cup B$ 30. $C = \{x \in U \mid -4 \leq x \leq -1\}$
31. $D = \{x \in U \mid -1 \leq x \leq 0\}$ 32. $C \cup D$

33. $C \cap D$ **34.** $\bar{C} \cap \bar{D}$

35. $K = \{(x, y) \in U \times U \mid y = 3x - 1\}$

36. $L = \{(x, y) \in U \times U \mid y = -2x + 4\}$

37. $K \cap L$

38. $P = \{(x, y) \in U \times U \mid y = -2x + 4\}$

39. $Q = \{(x, y) \in U \times U \mid y = 4x - 5\}$

40. $P \cap Q$

41. $R = \{(x, y) \in U \times U \mid y \geq x^2\}$

42. $S = \{(x, y) \in U \times U \mid y \leq 2x\}$

43. $R \cap S$ **44.** $R \cup S$

45. If the complement of the complement of A is denoted by $\bar{\bar{A}}$, show that $\bar{\bar{A}} = A$.

46. What is the complement of the universal set U?

47. What is the complement of the empty set ϕ?

In any triangle ABC, let L_1 be the set of all points on the bisector of angle A, let L_2 be the set of all points on the bisector of angle B, and let L_3 be the set of all points on the bisector of angle C.

48. Describe $L_1 \cap L_2$.

49. Describe $L_2 \cap L_3$.

50. Compare $(L_1 \cap L_2) \cap L_3$ and $L_1 \cap (L_2 \cap L_3)$.

58   Algebra of Sets

In algebra we have developed certain fundamental formulas which we use routinely to simplify expressions and to make calculations. Similarly, in the mathematics of sets there are certain formulas which enable us to manipulate two or more sets with directness and ease. The study of these formal manipulations of collections of sets is known as the *algebra of sets*. We shall see that in many respects, the algebra of sets is similar to ordinary algebra. We shall also note some important differences.

Commutative Laws

In ordinary algebra, the commutative laws for the operations of addition and multiplication are

$$a + b = b + a$$
$$ab = ba$$

In the algebra of sets, the commutative laws for the analogous operations of union and intersection are

$$A \cup B = B \cup A \tag{1}$$
$$A \cap B = B \cap A \tag{2}$$

The verification of Laws 1 and 2 follows directly from the definitions of the operations.   Venn diagrams may be used to illustrate these laws, as indicated in Figures 1 and 2.

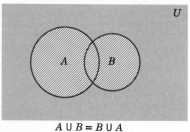

$A \cup B = B \cup A$

Figure 1

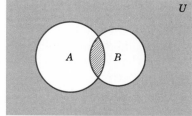

$A \cap B = B \cap A$

Figure 2

Associative Laws

The associative laws for the addition and multiplication of numbers are

$$(a + b) + c = a + (b + c)$$
$$(ab)c = a(bc)$$

The associative laws for the union and intersection of sets are similar in form.   They are

$$(A \cup B) \cup C = A \cup (B \cup C) \tag{3}$$
$$(A \cap B) \cap C = A \cap (B \cap C) \tag{4}$$

Again, the verification of these laws may be evidenced by Venn diagrams, as illustrated in Figures 3 and 4.

Distributive Laws

In the algebra of numbers we have noted that multiplication is distributive with respect to addition, that is,

$$a(b + c) = ab + ac$$

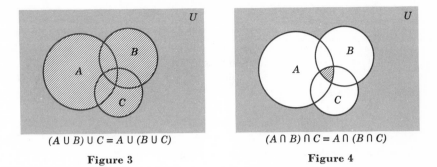

$(A \cup B) \cup C = A \cup (B \cup C)$

Figure 3

$(A \cap B) \cap C = A \cap (B \cap C)$

Figure 4

It is not true, however, that addition is distributive with respect to multiplication, and therefore the above is the only distributive law for numbers.

In the algebra of sets there is a dual distributive law as follows:

$$A \cup (B \cap C) = (A \cup B) \cap (A \cup C) \tag{5}$$
$$A \cap (B \cup C) = (A \cap B) \cup (A \cap C) \tag{6}$$

A diagrammatic representation of the distributive laws for sets may be facilitated by using separate Venn diagrams for each member of the equality. Thus in Figure 5, to illustrate Law 5, horizontal shading is

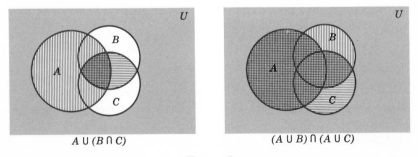

$A \cup (B \cap C)$

$(A \cup B) \cap (A \cup C)$

Figure 5

used to indicate $(B \cap C)$ in the first diagram, and vertical shading is used to indicate the set A. The set $A \cup (B \cap C)$ is the entire shaded area having horizontal, vertical, or double shading. In the second diagram of Figure 5, the vertical shading identifies the set $(A \cup B)$;

the horizontal shading identifies the set $(A \cup C)$. The set $(A \cup B) \cap (A \cup C)$ is therefore the area which is double shaded. It is apparent that the set $A \cup (B \cap C)$ in the first diagram is the same as the set $(A \cup B) \cap (A \cup C)$ in the second diagram which verifies Law 5. Law 6 may be verified in a similar manner.

Identity Laws

The following laws deal with operations on two sets which result in one of the sets. They are known as identity laws and are readily verified from the definitions of the symbols involved.

$$A \cup \phi = A \tag{7}$$
$$A \cap U = A \tag{8}$$
$$A \cup U = U \tag{9}$$
$$A \cap \phi = \phi \tag{10}$$

Law 7 resembles, in form, the law $a + 0 = a$ for addition of numbers and Laws 8 and 10 resemble, respectively, the laws $a \cdot 1 = a$ and $a \cdot 0 = 0$ for multiplication of numbers. Law 9 has no counterpart in the algebra of numbers.

Complement Laws

Certain important properties of complements are listed in the following formulas.

$$\bar{U} = \phi \tag{11}$$
$$\bar{\phi} = U \tag{12}$$
$$\bar{\bar{A}} = A \tag{13}$$
$$A \cup \bar{A} = U \tag{14}$$
$$A \cap \bar{A} = \phi \tag{15}$$

Laws 11 and 12 state that all the elements which are not in the universal set comprise the empty set and the converse. Statement 13 tells us that the complement of the complement of a set is the set itself. The last two formulas state that the union of a set with its complement is the universal set, and the intersection of a set with its complement is the empty set.

Additional Formulas

To complete our list of significant and useful formulas for an algebra of sets we add the following:

$$\overline{A \cup B} = \bar{A} \cap \bar{B} \tag{16}$$
$$\overline{A \cap B} = \bar{A} \cup \bar{B} \tag{17}$$
$$A \cup A = A \tag{18}$$
$$A \cap A = A \tag{19}$$

Formulas 16 and 17 are known as De Morgan's Laws. They are very useful in simplifying the complement of a set which is itself a composite of sets and operations. For example, the set $\overline{\bar{P} \cap Q}$ can be expressed directly in terms of its simple sets by an application of De Morgan's Laws. Thus using statement 17

$$\overline{\bar{P} \cap Q} = P \cup \bar{Q}$$

The form of the expression in the right member is both simpler and more useful than the form of the original expression. This will be more clearly apparent to the student who draws a Venn diagram to represent each member of the equality.

EXAMPLE 1.   Using any of the formulas 1 to 19, simplify the expression $P \cup (R \cup \bar{P})$.

$$\begin{aligned} P \cup (R \cup \bar{P}) &= P \cup (\bar{P} \cup R) & \text{(formula 1)} \\ &= (P \cup \bar{P}) \cup R & \text{(formula 3)} \\ &= U \cup R & \text{(formula 14)} \\ &= U & \text{(formula 9)} \end{aligned}$$

EXAMPLE 2.   Use the algebra of sets to simplify $(K \cap \bar{L}) \cup (K \cap L)$. Compare the given expression with the right member of formula 6:

$$A \cap (B \cup C) = (A \cap B) \cup (A \cap C)$$

Therefore as a consequence of this distributive law,

$$\begin{aligned} (K \cap \bar{L}) \cup (K \cap L) &= K \cap (\bar{L} \cup L) \\ &= K \cap U & \text{(formula 14)} \\ &= K & \text{(formula 8)} \end{aligned}$$

EXAMPLE 3.   Show by means of Venn diagrams that the simplification $(K \cap \bar{L}) \cup (K \cap L) = K$ holds for each of the following cases:

(a) K and L are disjoint
(b) $L \subset K$
(c) $K \subset L$
(d) $K = L$

In each diagram of Figure 6 the horizontal shading indicates the set $(K \cap \bar{L})$ and the vertical shading indicates the set $(K \cap L)$. The set $(K \cap \bar{L}) \cup (K \cap L)$ is therefore represented by all areas which are shaded. In each case the set so constituted is the set K.

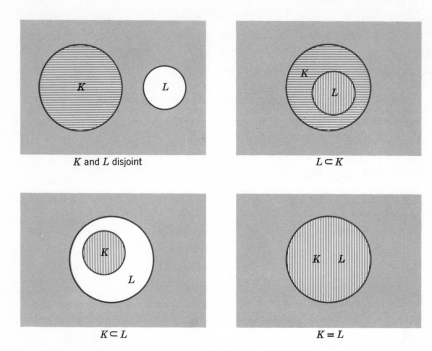

K and L disjoint   L ⊂ K

K ⊂ L   K = L

Figure 6

Duality

If, in any of the formulas 1 to 19 we interchange the symbols

$$\cup \text{ and } \cap$$
$$\phi \text{ and } U$$

the result is again one of the formulas. This is known as the *property of duality*. It illustrates one of the characteristics which gives the algebra of sets a simpler structure than the algebra of numbers.

EXAMPLE 4. Write the dual of the statement $A \cup \bar{A} = U$. If we replace the symbol $\cup$ by $\cap$, and U by ϕ, we have $A \cap \bar{A} = \phi$. Thus statement 15 of the list of formulas is the dual of statement 14.

Exercise 58

ALGEBRA OF SETS

Verify the commutative laws and De Morgan's Laws for each of the following pairs of sets by listing the elements of each member of the formulas:

1. $G = \{1, 2, 3, 4, 5\}$ and $K = \{-2, -1, 0, 1, 2, 3\}$.
   $U = \{-5, -4, -3, \ldots 5\}$
2. $B = \{1, 3, 5, 7, 9\}$ and $C = \{2, 4, 6, 8\}$.
   $U = \{0, 1, 2, 3, \ldots 9\}$
3. $R = \{a, b, c, d, e\}$ and $S = \{a, c, e\}$. $U = \{a, b, c, d, e, f\}$

Draw a Venn diagram to illustrate each of the following formulas:

4. $A \cup (B \cap C) = (A \cup B) \cap (A \cup C)$ 5. $A \cup \bar{A} = U$
6. $\bar{\bar{A}} = A$ 7. $\overline{A \cup B} = \bar{A} \cap \bar{B}$
8. $\overline{A \cap B} = \bar{A} \cup \bar{B}$

Express each of the following formulas verbally as a complete sentence:

9. $A \cup \phi = A$ 10. $A \cup U = U$
11. $A \cap U = A$ 12. $A \cap \phi = \phi$

What is the result of each operation (a) $A \cup B$ and (b) $A \cap B$ under the following conditions?

13. A is a proper subset of B.
14. B is a proper subset of A.
15. A and B are equal.

Using the principle of duality, write the dual of each of the following statements:

16. $A \cap B = B \cap A$ 17. $(A \cup B) \cup C = A \cup (B \cup C)$
18. $A \cap (B \cup C) = (A \cap B) \cup (A \cap C)$
19. $A \cup \phi = A$ 20. $A \cup U = U$
21. $\overline{A \cup B} = \bar{A} \cap \bar{B}$ 22. $A \cup A = A$

Using the algebra of sets, simplify each of the following expressions:

23. $P \cap \bar{P}$ 24. $R \cup \bar{\bar{R}}$
25. $\bar{A} \cap U$ 26. $\phi \cup \bar{G}$
27. $\bar{B} \cap \bar{U}$ 28. $P \cup (R \cup \bar{R})$

29. $(Q \cap P) \cap \bar{P}$

30. $(\bar{U} \cup B) \cup \bar{B}$

31. $G \cup (\bar{G} \cap H)$

32. $(K \cup N) \cap K$

33. $(B \cup C) \cap (B \cup \bar{C})$

34. $(F \cap G) \cup (G \cap \bar{G})$

35. $(R \cup \bar{S}) \cap (R \cup S)$

36. $(U \cup B) \cap (U \cup C)$

37. $(A \cap \phi) \cap (B \cap U)$

38. $(\bar{N} \cap L) \cup (\bar{N} \cap M)$

39. $(B \cap U) \cup (B \cup \phi) \cup \bar{\bar{B}}$

40. $\overline{\bar{R} \cup \bar{S}}$

41. $\overline{(B \cap \bar{C})} \cap C$

42. $(B \cup \phi) \cap (A \cup B)$

43. $P \cup \overline{\bar{R} \cup \bar{P}}$

44. $P \cap \overline{\bar{R} \cup \bar{P}}$

45. $P \cap [(\bar{P} \cap Q) \cup (\bar{Q} \cap P)]$

46. $P \cup [(P \cap \bar{Q}) \cup (\bar{P} \cap \bar{Q})]$

Draw a Venn diagram to show that $P \cap (P \cup Q) = P$ for each of the following cases:

47. $P \subset Q$

48. $Q \subset P$

49. $P = Q$

50. P and Q disjoint

Mathematical Tables

TABLE I.   POWERS AND ROOTS

SQUARES AND CUBES   SQUARE ROOTS AND CUBE ROOTS

No.	Square	Cube	Square Root	Cube Root	No.	Square	Cube	Square Root	Cube Root
1	1	1	1.000	1.000	51	2,601	132,651	7.141	3.708
2	4	8	1.414	1.260	52	2,704	140,608	7.211	3.733
3	9	27	1.732	1.442	53	2,809	148,877	7.280	3.756
4	16	64	2.000	1.587	54	2,916	157,464	7.348	3.780
5	25	125	2.236	1.710	55	3,025	166,375	7.416	3.803
6	36	216	2.449	1.817	56	3,136	175,616	7.483	3.826
7	49	343	2.646	1.913	57	3,249	185,193	7.550	3.849
8	64	512	2.828	2.000	58	3,364	195,112	7.616	3.871
9	81	729	3.000	2.080	59	3,481	205,379	7.681	3.893
10	100	1,000	3.162	2.154	60	3,600	216,000	7.746	3.915
11	121	1,331	3.317	2.224	61	3,721	226,981	7.810	3.936
12	144	1,728	3.464	2.289	62	3,844	238,328	7.874	3.958
13	169	2,197	3.606	2.351	63	3,969	250,047	7.937	3.979
14	196	2,744	3.742	2.410	64	4,096	262,144	8.000	4.000
15	225	3,375	3.873	2.466	65	4,225	274,625	8.062	4.021
16	256	4,096	4.000	2.520	66	4,356	287,496	8.124	4.041
17	289	4,913	4.123	2.571	67	4,489	300,763	8.185	4.062
18	324	5,832	4.243	2.621	68	4,624	314,432	8.246	4.082
19	361	6,859	4.359	2.668	69	4,761	328,509	8.307	4.102
20	400	8,000	4.472	2.714	70	4,900	343,000	8.367	4.121
21	441	9,261	4.583	2.759	71	5,041	357,911	8.426	4.141
22	484	10,648	4.690	2.802	72	5,184	373,248	8.485	4.160
23	529	12,167	4.796	2.844	73	5,329	389,017	8.544	4.179
24	576	13,824	4.899	2.884	74	5,476	405,224	8.602	4.198
25	625	15,625	5.000	2.924	75	5,625	421,875	8.660	4.217
26	676	17,576	5.099	2.962	76	5,776	438,976	8.718	4.236
27	729	19,683	5.196	3.000	77	5,929	456,533	8.775	4.254
28	784	21,952	5.292	3.037	78	6,084	474,552	8.832	4.273
29	841	24,389	5.385	3.072	79	6,241	493,039	8.888	4.291
30	900	27,000	5.477	3.107	80	6,400	512,000	8.944	4.309
31	961	29,791	5.568	3.141	81	6,561	531,441	9.000	4.327
32	1,024	32,768	5.657	3.175	82	6,724	551,368	9.055	4.344
33	1,089	35,937	5.745	3.208	83	6,889	571,787	9.110	4.362
34	1,156	39,304	5.831	3.240	84	7,056	592,704	9.165	4.380
35	1,225	42,875	5.916	3.271	85	7,225	614,125	9.220	4.397
36	1,296	46,656	6.000	3.302	86	7,396	636,056	9.274	4.414
37	1,369	50,653	6.083	3.332	87	7,569	658,503	9.327	4.431
38	1,444	54,872	6.164	3.362	88	7,744	681,472	9.381	4.448
39	1,521	59,319	6.245	3.391	89	7,921	704,969	9.434	4.465
40	1,600	64,000	6.325	3.420	90	8,100	729,000	9.487	4.481
41	1,681	68,921	6.403	3.448	91	8,281	753,571	9.539	4.498
42	1,764	74,088	6.481	3.476	92	8,464	778,688	9.592	4.514
43	1,849	79,507	6.557	3.503	93	8,649	804,357	9.644	4.531
44	1,936	85,184	6.633	3.530	94	8,836	830,584	9.695	4.547
45	2,025	91,125	6.708	3.557	95	9,025	857,375	9.747	4.563
46	2,116	97,336	6.782	3.583	96	9,216	884,736	9.798	4.579
47	2,209	103,823	6.856	3.609	97	9,409	912,673	9.849	4.595
48	2,304	110,592	6.928	3.634	98	9,604	941,192	9.899	4.610
49	2,401	117,649	7.000	3.659	99	9,801	970,299	9.950	4.626
50	2,500	125,000	7.071	3.684	100	10,000	1,000,000	10.000	4.642

From *General Mathematics*, by Currier, Watson, and Frame. Reprinted by permission of The Macmillan Company, publishers.

TABLE II.   COMMON LOGARITHMS

	0	1	2	3	4	5	6	7	8	9
10	0000	0043	0086	0128	0170	0212	0253	0294	0334	0374
11	0414	0453	0492	0531	0569	0607	0645	0682	0719	0755
12	0792	0828	0864	0899	0934	0969	1004	1038	1072	1106
13	1139	1173	1206	1239	1271	1303	1335	1367	1399	1430
14	1461	1492	1523	1553	1584	1614	1644	1673	1703	1732
15	1761	1790	1818	1847	1875	1903	1931	1959	1987	2014
16	2041	2068	2095	2122	2148	2175	2201	2227	2253	2279
17	2304	2330	2355	2380	2405	2430	2455	2480	2504	2529
18	2553	2577	2601	2625	2648	2672	2695	2718	2742	2765
19	2788	2810	2833	2856	2878	2900	2923	2945	2967	2989
20	3010	3032	3054	3075	3096	3118	3139	3160	3181	3201
21	3222	3243	3263	3284	3304	3324	3345	3365	3385	3404
22	3424	3444	3464	3483	3502	3522	3541	3560	3579	3598
23	3617	3636	3655	3674	3692	3711	3729	3747	3766	3784
24	3802	3820	3838	3856	3874	3892	3909	3927	3945	3962
25	3979	3997	4014	4031	4048	4065	4082	4099	4116	4133
26	4150	4166	4183	4200	4216	4232	4249	4265	4281	4298
27	4314	4330	4346	4362	4378	4393	4409	4425	4440	4456
28	4472	4487	4502	4518	4533	4548	4564	4579	4594	4609
29	4624	4639	4654	4669	4683	4698	4713	4728	4742	4757
30	4771	4786	4800	4814	4829	4843	4857	4871	4886	4900
31	4914	4928	4942	4955	4969	4983	4997	5011	5024	5038
32	5051	5065	5079	5092	5105	5119	5132	5145	5159	5172
33	5185	5198	5211	5224	5237	5250	5263	5276	5289	5302
34	5315	5328	5340	5353	5366	5378	5391	5403	5416	5428
35	5441	5453	5465	5478	5490	5502	5514	5527	5539	5551
36	5563	5575	5587	5599	5611	5623	5635	5647	5658	5670
37	5682	5694	5705	5717	5729	5740	5752	5763	5775	5786
38	5798	5809	5821	5832	5843	5855	5866	5877	5888	5899
39	5911	5922	5933	5944	5955	5966	5977	5988	5999	6010
40	6021	6031	6042	6053	6064	6075	6085	6096	6107	6117
41	6128	6138	6149	6160	6170	6180	6191	6201	6212	6222
42	6232	6243	6253	6263	6274	6284	6294	6304	6314	6325
43	6335	6345	6355	6365	6375	6385	6395	6405	6415	6425
44	6435	6444	6454	6464	6474	6484	6493	6503	6513	6522
45	6532	6542	6551	6561	6571	6580	6590	6599	6609	6618
46	6628	6637	6646	6656	6665	6675	6684	6693	6702	6712
47	6721	6730	6739	6749	6758	6767	6776	6785	6794	6803
48	6812	6821	6830	6839	6848	6857	6866	6875	6884	6893
49	6902	6911	6920	6928	6937	6946	6955	6964	6972	6981
50	6990	6998	7007	7016	7024	7033	7042	7050	7059	7067
51	7076	7084	7093	7101	7110	7118	7126	7135	7143	7152
52	7160	7168	7177	7185	7193	7202	7210	7218	7226	7235
53	7243	7251	7259	7267	7275	7284	7292	7300	7308	7316
54	7324	7332	7340	7348	7356	7364	7372	7380	7388	7396

From *College Algebra and Trigonometry*, by F. H. Miller. John Wiley and Sons (1945).

TABLE II.   COMMON LOGARITHMS
(Continued)

	0	1	2	3	4	5	6	7	8	9
55	7404	7412	7419	7427	7435	7443	7451	7459	7466	7474
56	7482	7490	7497	7505	7513	7520	7528	7536	7543	7551
57	7559	7566	7574	7582	7589	7597	7604	7612	7619	7627
58	7634	7642	7649	7657	7664	7672	7679	7686	7694	7701
59	7709	7716	7723	7731	7738	7745	7752	7760	7767	7774
60	7782	7789	7796	7803	7810	7818	7825	7832	7839	7846
61	7853	7860	7868	7875	7882	7889	7896	7903	7910	7917
62	7924	7931	7938	7945	7952	7959	7966	7973	7980	7987
63	7993	8000	8007	8014	8021	8028	8035	8041	8048	8055
64	8062	8069	8075	8082	8089	8096	8102	8109	8116	8122
65	8129	8136	8142	8149	8156	8162	8169	8176	8182	8189
66	8195	8202	8209	8215	8222	8228	8235	8241	8248	8254
67	8261	8267	8274	8280	8287	8293	8299	8306	8312	8319
68	8325	8331	8338	8344	8351	8357	8363	8370	8376	8382
69	8388	8395	8401	8407	8414	8420	8426	8432	8439	8445
70	8451	8457	8463	8470	8476	8482	8488	8494	8500	8506
71	8513	8519	8525	8531	8537	8543	8549	8555	8561	8567
72	8573	8579	8585	8591	8597	8603	8609	8615	8621	8627
73	8633	8639	8645	8651	8657	8663	8669	8675	8681	8686
74	8692	8698	8704	8710	8716	8722	8727	8733	8739	8745
75	8751	8756	8762	8768	8774	8779	8785	8791	8797	8802
76	8808	8814	8820	8825	8831	8837	8842	8848	8854	8859
77	8865	8871	8876	8882	8887	8893	8899	8904	8910	8915
78	8921	8927	8932	8938	8943	8949	8954	8960	8965	8971
79	8976	8982	8987	8993	8998	9004	9009	9015	9020	9025
80	9031	9036	9042	9047	9053	9058	9063	9069	9074	9079
81	9085	9090	9096	9101	9106	9112	9117	9122	9128	9133
82	9138	9143	9149	9154	9159	9165	9170	9175	9180	9186
83	9191	9196	9201	9206	9212	9217	9222	9227	9232	9238
84	9243	9248	9253	9258	9263	9269	9274	9279	9284	9289
85	9294	9299	9304	9309	9315	9320	9325	9330	9335	9340
86	9345	9350	9355	9360	9365	9370	9375	9380	9385	9390
87	9395	9400	9405	9410	9415	9420	9425	9430	9435	9440
88	9445	9450	9455	9460	9465	9469	9474	9479	9484	9489
89	9494	9499	9504	9509	9513	9518	9523	9528	9533	9538
90	9542	9547	9552	9557	9562	9566	9571	9576	9581	9586
91	9590	9595	9600	9605	9609	9614	9619	9624	9628	9633
92	9638	9643	9647	9652	9657	9661	9666	9671	9675	9680
93	9685	9689	9694	9699	9703	9708	9713	9717	9722	9727
94	9731	9736	9741	9745	9750	9754	9759	9763	9768	9773
95	9777	9782	9786	9791	9795	9800	9805	9809	9814	9818
96	9823	9827	9832	9836	9841	9845	9850	9854	9859	9863
97	9868	9872	9877	9881	9886	9890	9894	9899	9903	9908
98	9912	9917	9921	9926	9930	9934	9939	9943	9948	9952
99	9956	9961	9965	9969	9974	9978	9983	9987	9991	9996

Answers to Odd-Numbered Exercises

Exercise 1

1. 3, 6, 9, 12, 15, 18, 21, 24, 27.   3. Empty set.   5. Empty set.
7. Does not.   9. Does not.   11. Does.   13. Does not.
15. Does not.   17. Does.   19. Is.   21. Is not.
23. Is not.   25. Is not.   27. Is.   29. Is not.
31. Has.   33. Has not.   35. Has not.   37. 26.
39. 12.   41. 28.   43. 8.   45. 4.   47. 14.
49. A starting point is lacking in each case, resulting either in repeated referral to undefined terms or in circuitous sequences of definitions.
51. True.   53. True.   55. False.   57. Infinite.
59. Infinite.   63. The set of perfect squares between 36 and 81, inclusive. (Or an equivalent statement.)   65. The set of natural numbers less than 40 which end in 6. (Or an equivalent statement.)   67. Quadrilaterals, regular polygons, isosceles triangles, etc.   69. Beetles (*Coleoptera*), butterflies and moths (*Lepidoptera*), termites (*Isoptera*), etc.   71. 2, 6, 8, 10.   73. 2, 4, 6, 8.   75. 2, 4.
77. 6.   79. 4, 6, 8.

Exercise 2

1. Multiplication.   3. Addition.   5. Subtraction.
7. Multiplication.   9. Root.   11. Root.   13. Power.
15. Impossible.   17. Impossible.   19. Possible.
21. Possible.   23. Possible.   25. Impossible.   27. Possible.
29. Impossible. (Zero is not a natural number.)   31. Impossible.
33. Impossible.   35. Root.   37. Base.   39. Exponent.
41. Difference.   43. $24 - 4 = 20$; $20 - 4 = 16$; etc.   45. 4.
47. 16.   49. 36.   51. 27.   53. 125.   55. 16.   57. 256.
59. 1296.   61. 2.   63. 2.   65. 3.   67. 6.   69. 3.
71. 4.   73. 4.   75. $4 \cdot 3$.   77. $3 \cdot 5$.   79. 13^6.
81. $q - 4 = p$; $q - p = 4$.   83. $p + p + p + p + p = q$.   85. 100.
87. 4.

Exercise 3

1. $7 - 2 \neq 2 - 7$; $7 - (4 - 1) \neq (7 - 4) - 1$; etc.   3. Commutative law of multiplication.   5. Associative law of addition.   7. Associative law of multiplication.   9. Associative law of addition.

11. $a(bc) = (ab)c$ (associative law of multiplication)
    $= c(ab)$ (closure law, commutative law of multiplication).
13. $a + (b + c) = (a + b) + c$ (associative law of addition)
    $= c + (a + b)$ (closure law, commutative law of addition).
15. $(3x + 3y)z = 3(x + y)z$ (distributive law)
    $= 3z(x + y)$ (closure law, commutative law of mult.).
17. $(a + 2)(b + c) = (a + 2)b + (a + 2)c$ (closure law of addition, distributive law)
    $= b(a + 2) + c(a + 2)$ (closure law of addition, commutative law of mult.)
    $= ba + b \cdot 2 + ca + c \cdot 2$ (distributive law)
    $= ab + 2b + ac + 2c$ (commutative law of mult.).
19. Closed.   21. Not closed.   23. Not closed.   25. Not closed.
27. Not commutative.   29. Not commutative.   31. Not commutative.
33. (a) If a and b are sisters, then a is the sister of b, and b is the sister of a.
    If a and b play on the same team, then a is a teammate of b and b is a teammate of a.
    If a is an employer and b is his employee, then a is an employer of b, but b is not an employer of a.
    If a is the father of b, then b is not the father of a.
    Similarly for part (b).

Exercise 4

1. $E > B$.   3. $G > E$.   5. $I < J$.   7. $D < F$.   9. No.
11. Yes.   13. Yes.   15. No.   17. No.   19. $+3$.
21. $+50$.   23. $+200$.   25. $+3°$.   27. $+7°$.   29. 10.
31. 14.   33. 11.   35. 7.   37. 32,   39. 4.   41. 5.

Exercise 5

1. 13.   3. -1.   5. -6.   7. 0.   9. 0.   11. -6.
13. 0.   15. -4.   17. (a) 2; (b) 2.   19. (a) 6; (b) 6.
21. 4.   23. -7.   25. -3.   27. -12.   29. 12.   31. -8.
33. 60.   35. -11.   37. 0.   39. 11.   41. -15.   43. -21.
45. -27.   47. (a) -9; (b) -9; (c) associative law.   49. -40.
51. -40.   53. -36.   55. 21.   57. 48.   59. 36.
61. -96.   63. 8.   65. 32.   67. 4.   69. 16.   71. 64.
73. (a) -60; (b) 60; (c) 30; (d) -60.

Exercise 6

1. 6.   3. -5.   5. 2.   7. 10.   9. 10.   11. 11.
13. -3.   15. -23.   17. -5.   19. 18.   21. 5.   23. -6.
25. 3.   27. 9.   29. 25.   31. 23.   33. 34.   35. 2.
37. -14.   39. -12.   41. -23.   43. -3.   45. -3.   47. 2.
49. -10.   51. 5.   53. 4.   55. -3.   57. -3.   59. 1.

61. -2.   63. 4.   65. 2.   67. -3.   69. -4.   71. -2.
73. 3.   75. 2.   77. -3.   79. -2.   81. 8, 4, 12, 3.
83. -8, -4, 12, 3.   85. 24, 8, 128, 2.   87. -8, -24, -128, -2.
89. -6, -6, 0, not defined.   91. -4, 4, 0, 0.   93. -6, 6, 0, 0.
95. Has no meaning.   97. Has meaning.   99. Has meaning.
101. Has meaning.   103. Has meaning.   105. Has meaning.

Exercise 7

3. C is the circumference of a circle, r is its radius.   5. A is the area of a circle, r is its radius.   7. i is the amount of interest earned, p is the principal, r is the rate per period, t is the number of periods.   9. V is the volume of a rectangular solid, l is its length, w is its width, h is its height.   11. V is the volume of a circular cylinder, r is its radius, h is its height.   13. a is the average of a set of numbers, s is their sum, n is the number of elements.   15. p is the perimeter of an equilateral quadrilateral, s is the length of one side.   17. p is the perimeter of a rectangle, l is its length, w is its width.   19. s is the distance in feet covered by a freely falling body, t is the number of seconds it is falling.
21. $365 \leq n \leq 366$.   34. $4 < l < 10$.   25. $z < 21$.
29. $6 \leq v \leq 31$.   31. $4 + 4 + 4 = 3 \cdot 4$, etc.
33. $7 \cdot 19 + 19 = 8 \cdot 19$, etc.   35. $3^2 \cdot 5^2 = (3 \cdot 5)^2$, etc.
37. Variable.   39. Variable.   41. Variable.

Exercise 8

1. Binomial.   3. Binomial.   5. Monomial.   7. Monomial.
9. Trinomial.   11. Monomial.   13. Monomial.   15. Binomial.
17. $-2av$.   19. $7y^2$.   21. b/ax.   23. $2\pi rh$.   25. 3, $3y$.
27. $-\frac{5}{2}$, $-\frac{5}{2}k$.   29. 1, 1.   31. $\frac{1}{4}$, $\frac{1}{4}$.   33. 1, y/z.   35. 2, $2yz$.
37. $x^2 + x - 5$, 2.   39. $-16l^2 + 4$, 2.   41. $x^n + x^{n-1} - 5x^{n-3} - 3$, n.
43. $4v^3 + 3v^2 - 7v + 8$, 3.   45. $a^6 + a^4 + a^2 + 1$, 6.   47. 2, 2, 2.
49. 2, 1, 3.   51. 1, 3, 4.   53. Polynomial in x.
55. Not a polynomial in x.   57. Polynomial in x.   59. Polynomial in x.
61. Not a polynomial in x.   63. $7x^2y$ $\sqrt{2}\, x^2y$

Exercise 9

1. $\frac{1}{4}$.   3. $-\frac{1}{6}$.   5. $1/x$.   7. $-8/p$.   9. -7.   11. $+9$.
13. Yes, since every integer can be written as a fraction with denominator equal to 1.
15. $\frac{7}{6}$.   17. $\frac{14}{15}$.   19. $-\frac{1}{3}$.   21. $\frac{37}{22}$.   23. $-\frac{2}{5}$.   25. $\frac{9}{20}$.
27. Postulates 8, 9, 11.   29. (a) 2; (b) no, because between any two rational numbers there is another rational number.

Exercise 10

1. 8.   3. 100.   5. -30.   7. $\frac{9}{5}$.   9. $\frac{41}{37}$.   11. -14.
13. 108.   15. 14.   17. -40.   19. -20.   21. 4.   23. 14.
25. 99.   27. -8.

29. For $x = 3$, $28 = 28$; for $x = 5$, $34 \neq 40$; for $x = 1$, $22 \neq 16$.

31. For $y = -1$, $0 = 0$; for $y = 1$, $-10 \neq 0$; for $y = -2$, $32 \neq 0$.

33. For $b = -3$, $\frac{7}{12} = \frac{7}{12}$; for $b = 2$, $6 \neq -\frac{3}{2}$; for $b = 3$, $\frac{13}{6} \neq -\frac{5}{6}$.

35. For $z = -\frac{5}{2}$, $-\frac{5}{2} = -\frac{5}{2}$; for $z = 1$, $-\frac{3}{4} \neq \frac{13}{6}$; for $z = \frac{5}{2}$, $0 \neq \frac{50}{12}$.

37. For $x = -\frac{1}{2}$, $16 = 16$; for $x = \frac{1}{2}$, $4 \neq 14$; for $x = \frac{1}{3}$, $\frac{49}{9} \neq \frac{124}{9}$.

43. Increasing.   45. Increasing.   47. Both.   49. Both.

51. Increasing.   53. 325.   55. 144.   57. $12 \cdot 6$.   59. 35.

61. 7900.   63. $c(x + y)$.   65. $\dfrac{4(7a + 3b)}{a + b}$.

Exercise 11

1. $4x^4$.   3. $-12ab^2c$.   5. $2r^2t^2s$.   7. $2xxyy$.   9. $-3pppqqqq$.

11. $-uuv$.   15. $-14x^2y^5z^6$.   19. $-24m^2n^3$.   21. b^2cd^3.

23. r^5.   25. $-15y^3$.   27. $24w^6$.   29. 5^6.   31. 4^8.

33. $6a^4$.   35. $21a^3xy^4$.   37. $-16mn^3x^4$.   39. $66d^4m^2x^5y$.

41. x^5y^6.   43. $-30abc^3d$.   45. $8n^4$.   47. x^2y^4.   51. 3.

53. -2.   55. x.   57. -1.   59. -1.   61. a.   63. $-7u$.

65. $-3v$.   67. $-b$.   69. $7a^2b^2x$.   71. $17x^5y^2z$.   73. 0.

75. $8(a + b)$.   77. $-2(2c - d)$.   79. $-4ab$.   81. $11y^3$.

83. $15(a - 2b)$.   85. $-(p + q)$.   87. x.   89. $-2mn$.

91. 0.   93. $3x$.   95. $-a^2m^2x$.   97. $8b$.   99. $-10ab$.

101. $4b^8$.   103. $27v^9$.   105. $25x^{10}$.   107. $-216b^3x^9$.

109. $-729a^3b^6c^3$.   111. $-1000x^9y^3z^{12}$.   113. a^2.   115. $-27b^9$.

117. $7a^3b^2$.   119. $12b^4x^7y^5$.   121. t^2.   123. $5a^3x$.   125. rs^4.

127. $1.3a^2b$.

Exercise 12

1. $5x + 1$.   3. $5v - 5w$.   5. $-18y^2$.   7. $12a + 10b + 3c$.

9. $7x^2 - 5x + 4$.   11. $10a^2 - 6b^2 - 5bc + 4c^2$.

13. $3x^2 - 12xy - y^2$.   15. $14p - 6q - 13r - 3$.

17. $6m^2n^2 + 11m^2n + 2mn^2 - 7mn$.   19. $10xy - 6x$.   21. 16.

23. $8x^3 - 7x^2 + 8x - 5$.   25. $2y$.   27. $2x^2 + x + 6$.

29. $-4x^3 + 12x^2$.   31. $4b$.   33. $-x - 2y$.   35. $17a^2 + 36ab + 31a$.

37. $53x$.   39. $-2x + 4z$.   41. $4a^2 - 20a + 25$.

43. $9x^2 - 6xy + y^2$.   45. $4p^2 + 36pq + 56q^2$.   47. $49x^2 - 16$.

49. $-25b^2 + 35b - 6$.   51. $8a^3 - 10a^2 + 5a - 12$.

53. $9x^5 + 3x^4 - 21x^3 + x^2 + 10x - 2$.   55. $a^2 - a$.

57. $-a + b - c$.   59. $-x + 9x^3y$.   61. $2ab - 6a^3b - 4a^2 + 3$.

63. $-x^2 + x - 1$.   65. $x - 2$.   67. None.   69. None.

71. $c^2 - 2c - 1$, Rem. $= -3$.   73. $x^2 + 2x + 3$, Rem. $= 7$.

75. $2x + h - 1$.   77. $4x + 2h - 7$.

Exercise 13

9. $x = 0$.   11. None.   13. $a = 2$.   25. Equation.

27. Equation.   29. Identity.   31. Equation.   33. Identity.

35. Equation.   37. Equation.   39. Identity.   41. Equation.

43. $9x - 6 = 30$.   45. $3x = 12$.   55. Satisfied.

57. Not satisfied.   59. Not satisfied.   61. Not satisfied.

63. Equation, since it is not true for all admissible values of C and r.

Exercise 14

1. Symmetric property.   3. Addition or subtraction property.

5. Reflexive property.   7. Multiplication property.

9. Addition property.   11. Symmetric property.   13. Symmetric property.

15. Reflexive property; transitive property.

17. x.   19. b.   21. x.   23. Add 2.   25. Add -5.

27. Divide by -6.   29. Multiply by $\frac{3}{2}$.   31. Multiply by $-\frac{4}{3}$.

33. Divide by -2.9.   35. Multiply by 8.   37. 8.   39. -11.

41. -37.   43. 15.   45. -9.   47. $\frac{3}{2}$.   49. -20.

51. 55.   53. -15.   55. 100.   57. -24.   59. -7.

61. 3.   63. 6.   65. 5.   67. 6.   69. -1.   71. 2.

73. -5.   75. 2.   77. 1.   79. $-\frac{4}{3}$.   81. 2.   83. $\frac{6}{7}$.

85. 21.   87. $-\frac{1}{2}$.   89. -7.   91. $\frac{1}{3}$.   93. 26.

95. -24.   97. $\frac{7}{6}$.   99. $\frac{31}{3}$.   101. (a) 20 ohms; (b) 3 amperes;

(c) 2 volts.   103. 20 feet per second per second.   105. (a) $4\frac{1}{2}$ hours; (b) 1120 centimeters per second.   107. (a) 8 inches; (b) 5 feet.   109. (a) 12 square feet; (b) $\frac{45}{7}$ inches.

Exercise 15

1. $x + 6$.   3. $m + n$.   5. $u - c$.   7. $y + 0.1y$.   9. $a + n$.

11. $17 - x$.   13. $14x$.   15. $60h$.   17. $4m + 5n$.   19. $3(n - 3)$.

21. $3r$.   23. $n/60h$.   25. $6w + 8$.   27. (a) $(24 - x) - x = 18$;

(b) $(24 - x) + 4x = 45$; (c) $(24 - x) - x = x - 9$.

29. (a) $10d + 25(22 - d) = 385$; (b) $d = (22 - d) + 8$;

(c) $10d = 25(22 - d) + 170$.

Exercise 16

1. $5\frac{1}{2}$ feet, $9\frac{1}{2}$ feet.   3. 27, 29.   5. 27°, 72°, 81°.

7. 17.   9. 14, 42.   11. 17 dimes, 14 quarters.   13. 23 years, 28 years.

15. 6 years, 12 years.   17. 22 feet, 37 feet.   19. $1\frac{1}{5}$ hours.

21. 11 inches, 17 inches, 22 inches.   23. 19.   25. 7.

27. $3500 at 5%, $2500 at 4%.   29. 3740 feet.

Exercise 17

1. $2a > 2b$.   3. $a - 2 > b - 2$.   5. $-a < -b$.   7. $2 - a < 2 - b$.

17. False, $x > -6$ by Postulate 3d.   19. True, by Postulate 3d.

21. Conditional.   23. Conditional.   25. Absolute.   27. Absolute.

29. Conditional.   31. $x < 5$.   33. $x < 7$.   35. $x \geq 6$.

37. $x > -6$.   39. $x < 30$.   41. $x \leq -3$.   43. $4 < x < 6$.

45. $4 \leq x \leq 6$.   47. $x > 5$, and $x < -5$.

Exercise 18

1. $x = 6, y = 1.$ 3. $x = -2, y = -3.$ 5. $x = 12, y = -9.$
7. $x = -\frac{5}{8}, y = \frac{23}{8}.$ 9. $x = \frac{5}{14}, y = -\frac{3}{14}.$
11. $x = 1, y = 1.$ 13. $x = 5, y = 4.$ 15. $x = 7, y = 9.$
17. $x = \frac{57}{8}, y = \frac{21}{8}.$ 19. $a = 1, b = 1.$ 21. $u = \frac{54}{13}, v = -\frac{18}{13}.$
23. $b = -5, c = -3.$ 25. $p = 1, q = 1.$ 27. $d = -2, t = 6.$
29. No solution.   31. One solution.   33. One solution.   35. No solution.

Exercise 19

1. 127, 83.   3. 37, 12.   5. 126, 18.   7. $a = -4, b = 7.$
9. 82.   11. 18 nickels, 7 quarters.   13. 54.   15. 140 quarters,
100 dimes.   17. 18, 8.   19. 124 feet, 40 feet.   21. 50 gallons of 90% solution,
150 gallons of 70% solution.   23. $8500 at 5%, $3500 at 6%.   25. 12 quarts.

Exercise 20

1. $6x^2 + 17xy + 12y^2.$ 3. $6x^2 - 13xy + 6y^2.$ 5. $4t^2 + 8t - 21.$
7. $x^2 + 10x + 21.$ 9. $x^2 + 4x - 21.$ 11. $a^2 + 3ab + 2b^2.$
13. $3x^2 - 5xy + 2y^2.$ 15. $3x^2 + 5xy + 2y^2.$ 17. $9x^2 - 24x + 16.$
19. $4t^2 - 28t + 49.$ 21. $16 + 24x + 9x^2.$ 23. $\pi^2 + 2\pi h + h^2.$
25. $r^2 + 2rr' + (r')^2.$ 27. $b_1^2 + 2b_1b_2 + b_2^2.$ 29. $4x^2 - 49.$
31. $t^4 - 36.$ 33. $r^2 - (r')^2.$ 35. $y^4 - 9.$ 37. 1225.
39. 441.   41. 256.   43. 3969.   45. 160, 801.
47. 636, 804.   49. 384.   51. 899.   53. 399.   55. 1596.
57. $18xy.$ 59. 49.   61. $4t.$ 63. $40m.$ 65. $9x^2.$
67. $x^2 + 2xy + y^2 - 4.$ 69. $x^2 + 2xy + y^2 + 5x + 5y + 6.$
71. $4s^2 - t^2 - 6t - 9.$ 73. $9x^2 - 4y^2 + 4y - 1.$ 75. $n^4 - 16.$
77. $n^4 - 8n^2 + 16.$

Exercise 21

1. $(x + 2)(x + 1).$ 3. $(2n + 3)(n + 2).$ 5. $(3x + 2)(x + 2).$
7. $(x + 8y)(x - 3y).$ 9. $(x - 7)(x + 4).$ 11. $(2x + z)(x + 3z).$
13. $(2x + 1)(x - 2).$ 15. $(2x + 3y)(4x - y).$ 17. $(7x + 1)(x - 6).$
19. $(2x + 3)(x - 2).$ 21. $(x - 43)(x + 1).$ 23. $(t - 6)(t + 5).$
25. $(a + 2)^2.$ 27. $(4t + 1)^2.$ 29. $(4t - a)^2.$ 31. $(5x - 2)^2.$
33. $(13x + 3)^2.$ 35. $(3 - a)^2.$ 37. $(x - 10)^2.$
39. $(2m + 1)^2.$ 41. $(a - 1)(a + 1).$ 43. $(2t - 5r)(2t + 5r).$
45. $(6cd - 3)(6cd + 3).$ 47. $(1.5a - 1.3b)(1.5a + 1.3b).$
49. $a^2(a + 1).$ 51. $3(x - 2).$ 53. $3a(x - 2).$
55. $2\pi(RH + rh).$ 57. $\pi r(r + 2).$ 59. $3c(c - 4 - 6c^3).$
61. $5(c^2 + 2c - 3)$ or $5(c + 3)(c - 1).$ 63. $3xy(2x^2 + y - 3xy).$
65. $2a^2(1 + a + a^2).$ 67. Is not.   69. Is.   71. Is not.
73. Is. 75. Is. 77. The factors will consist of the common factor and the
expression obtained by dividing each term of the given expression by the common
factor.   79. $(a + t + 2)(a + t - 2).$ 81. $(2x - 1)(2x - 11).$
83. $(x + 1 + t)(x + 1 - t).$ 85. $(x + a - 1)(x - a + 1).$

Exercise 22

1. $3(x-2)^2$.   3. $4(a-6)(a+6)$.   5. $2(3x+2)(x+5)$.
7. $2(x-2y)(x+2y)$.   9. $2n(n-1)(n+1)$.
11. $2(x^4+1)(x^2+1)(x+1)(x-1)$.   13. $(4n^2+3)(n^2-3)$.
15. $5n^4(a-4n)$.   17. $(x^2-2)^2$.   19. $5x(x-2y)(x+2y)$.
21. $2a(x+7)^2$.   23. $6(4x-3y)(x+y)$.
25. $5(x^2+4)(x+2)(x-2)$.   27. $2(2x-3)(2x+3)(x-1)(x+1)$.
29. $m(m-6)(m+5)$.   31. $P=2(l+w)$.   33. $S=\pi r(r+s)$.
35. $S=n(n+1)$.   37. $A=\frac{1}{2}h(b+b')$.   39. $A=P(1+rt)$.
43. $(x+y-1)(x-y+1)$.   45. $(x+y-z)(x+y+z)$.
47. $(a+b+m+n)(a+b-m-n)$.   49. $(2x+y+2)(2x-y)$.

Exercise 23

1. Is.   3. Is not.   5. Is.   7. Is.   9. Is.
11. $2x^2+3x-6=0$.   13. $x^2-3x-17=0$.   15. $x^2-2x-19=0$.
17. $2x^2+6x-13=0$.   19. $x^2+12x-17=0$.   21. $x^2+17x-12=0$.
23. $x^2-6x+6=0$.   25. $a=2, b=-7, c=7$.   27. $a=1,$
$b=3, c=-4$.   29. $a=3, b=-4, c=-4$.   31. $a=4, b=3, c=-7$.
33. $a=a, b=-d, c=k$.   35. -1.   37. $5, -5$.   39. $2, -3$.
41. $3, \frac{2}{7}$.   43. None.   45. $3, -3$.   47. $3, -3$.   49. $2, -2$.
51. Does.   53. Does.   55. Does.   57. 0.   59. $6x^2-9x-1=0$.
61. $5x^2-x-1=0$.   63. $x^2-2x-15=0$.   65. $x^2-2x+\frac{7}{2}=0$.
67. $x^2-\frac{3}{5}x-\frac{12}{5}=0$.   69. $x^2-\frac{1}{2}x-25=0$.

Exercise 24

1. $3, -3$.   3. $0, 7$.   5. $7, -6$.   7. $1, 4$.   9. $15, -2$.
11. $4, 4$.   13. $1, 2$.   15. $\frac{5}{2}, -3$.   17. $\frac{9}{2}, -\frac{4}{3}$.
19. $\frac{1}{2}, \frac{5}{3}$.   21. $\frac{3}{4}, -2$.   23. $2a, -3a/2$.   25. $0, 1, 2$.
27. $0, 4, -4$.   29. $0, \frac{2}{3}, \frac{3}{2}$.   31. $0, 2, 3$.   33. $0, 2, \frac{3}{2}$.
35. $x^2-5x+6=0$.   37. $x^2-9=0$.   39. $2x^2-x=0$.
41. $6x^2+x-2=0$.   43. $9x^2-4=0$.   45. $x^2+4x+4=0$.
47. $13\frac{1}{2}$ seconds.   49. 5 seconds.   51. 10 feet.

Exercise 25

1. Is.   3. Is.   5. Is.   7. Is.   9. Is.   11. Is not.
13. Is not.   15. Is.   17. $2\sqrt{6}$.   19. $6\sqrt{2}$.   21. $5\sqrt{2}$.
23. $6\sqrt{7}$.   25. $8\sqrt{3}$.   27. $8\sqrt{5}$.   29. $2\sqrt[3]{2}$.   31. $6\sqrt[3]{2}$.
33. $10\sqrt[3]{5}$.   35. $2\sqrt[3]{4}$.   37. $\sqrt{2}/2$.   39. $\sqrt{5}/5$.   41. $\sqrt{6}/4$.
43. $\sqrt{30}/6$.   45. $\sqrt[3]{2}/2$.   47. $\sqrt[3]{4}/4$.   49. $\sqrt[3]{6}/2$.
51. $\sqrt[3]{50}/5$.   53. $\sqrt{2}+8\sqrt{3}-6$.   55. $23\sqrt{2}$.   57. $22\sqrt{3}$.
59. $7\sqrt{5}-8\sqrt{2}$.   61. $\sqrt[3]{2}$.   63. $\sqrt{6}$.   65. $3\sqrt{2}$.   67. 7.
69. $8\sqrt{26}$.   71. 24.   73. 7.   75. $3+2\sqrt{3}+\sqrt{6}$.
77. $\sqrt{10}-5\sqrt{2}-10$.   79. -1.   81. -8.   83. $7-2\sqrt{10}$.

85. 8.   87. $31 - 4\sqrt{21}$.   89. $\sqrt{2}$.   91. $\sqrt{6}/2$.   93. $\sqrt{15}/2$.

95. $5\sqrt{5}/8$.   97. $\sqrt{2}$.   99. $\sqrt{2}/2$.   101. 3.   103. $\sqrt[3]{4}$.

105. $2 - \sqrt{3}$.   107. $2 - \sqrt{6}$.   109. $\dfrac{3 + \sqrt{2}}{3}$.   111. $\frac{1}{2} + \sqrt{2}$.

113. 1.414.   115. 1.155.   117. 0.816.   119. 1.732.

121. 3.414.   123. 2.207.   125. -0.674.   127. 0.382.

129. -0.581.

Exercise 26

5. U, R, I, W.   7. U, R, I, W.   9. U, R.

11. U, R.   13. $\frac{7}{16} = 0.4375$.   15. $\frac{31}{4} = 7.75$.

17. $1.78 = \frac{89}{50}$.   19. $0.1313\ldots = \frac{13}{99}$.   21. $0.003 = \frac{3}{1000}$.

23. Irrational.   25. Irrational.   27. Rational.

29. Commutative postulate.   31. Commutative postulate.

33. Associative postulate.

Exercise 27

1. $1, \frac{3}{2}$.   3. $4, \frac{1}{2}$.   5. $-\frac{2}{3}, -\frac{5}{2}$.   7. $-2, \frac{3}{4}$.

9. $2, -\frac{5}{2}$.   11. $\frac{3}{2}, -\frac{1}{2}$.   13. $\frac{1}{3}, \frac{2}{3}$.   15. $-4, \frac{2}{5}$.

17. $2, -3$.   19. $\frac{1}{2}, \frac{2}{5}$.   21. $0.366, -1.366$.   23. $1.186, -1.686$.

25. $0.419, 3.581$.   27. $-3.414, -0.586$.   29. $0.191, 1.309$.

31. Real, unequal, irrational.   33. Complex.   35. Complex.

37. Real, equal, rational.   39. Real, unequal, irrational.

Exercise 28

1. 0 or $\frac{25}{4}$.   3. 5 or -8.   5. 19, 20 or $-20, -19$.

7. 20 feet by 27 feet.   9. 8 inches, 15 inches.   11. 6 inches, 8 inches

13. 0.26 inch.   15. $2\frac{1}{4}$ seconds.   17. $6\frac{1}{4}$ seconds.

19. $156\frac{1}{4}$ feet.   21. 48 inches.   23. 0.125 mile.   25. 8.090 feet.

Exercise 29

1. $x = 2$.   3. $x = b$.   5. None.   7. $x = 1$.   9. None.

11. $2x$.   13. x^2z.   15. cr^2.   17. $x^2 - y^2$.   19. $xy - y^2$.

21. $x - y$.   23. $2/5x$.   25. $x/7a$.   27. $6x$.   29. a/b^3.

31. $3x/4y$.   33. $3/(n - 3)$.   35. $(x - 1)/(x + 1)$.

37. $(x + a)/a^3$.   39. $(x + 4)/(x + 5)$.   41. $a/(x + y)$.

43. $(2y - 5)/(y - 4)$.   45. $(x - 2)/2(x - 3)$.   47. Plus.

49. Plus.   51. Plus.   53. $\frac{1}{2}$.   55. Not possible.

57. -1.   59. $-a/2$.   61. $-n - 2$.   63. $-\dfrac{x + 3}{y}$.

65. Not possible.   67. Not possible.   69. Not possible.

71. No, because $(x - 2)^2 = (2 - x)^2$.   73. $\dfrac{(2 - x)^2}{x + 1}$.

Exercise 30

1. $8y/9$.   3. $1/x^3$.   5. $b/3ac$.   7. $10/3a^3$.   9. $7xy/4$.

11. $x + y$.   13. $\dfrac{ab(3a + b)}{3a - b}$.   15. $(x + 2)/2$.   17. $5c/(a + b)$.

19. $(x + 2)/(x - 2)$.   21. x^2.   23. -5.   25. $5/3$.

27. $2/x$.   29. $(x - 2)/2$.   31. $(a + c)/(a + b)$.

33. $\dfrac{a^2 - 1}{a^2 - 2a + 8}$.   35. $a/2n$.   37. $3xy/2$.   39. $x/14y$.

41. $2/(a - b)$.   43. $-\dfrac{a + b}{a + 2b}$.   45. $\dfrac{2(x - 2)}{(x + 2)(x - 1)}$.

47. $(x - 3)(4 - 3x)$.

Exercise 31

1. $2^3 \cdot 3^2$.   3. $3 \cdot 5 \cdot 7ax^2$.   5. $2^2\pi r^2 h$.   7. $2^2 \cdot 3^2 axyz$.
9. $2^2 \cdot 3^2 \cdot 7x^2y^2$.   11. $2^2x(x - 2)$.   13. $2^2(x + 2y)$.
15. $2^2x(x - 1)(x - 2)$.   17. $2x(x - 1)^2(x + 1)$.
19. $y(x - 2y)(x + 2y)$.   21. $7x/5$.   23. $(15x + 8x^2)/20$.
25. $(5x - 18)/12$.   27. $(x - 33)/30$.   29. $19/5x$.

31. $(10x + 12x^2)/15$.   33. $\dfrac{5x^3 + 6x + 14}{10x^2}$.   35. $\dfrac{2x - 6y}{3xy}$.

37. $\dfrac{2x - 1}{x - 1}$.   39. $\dfrac{4a^3 - 3a^2 + 3}{a(a^2 - 1)}$.   41. $\dfrac{x^2 + 3x - 1}{(x - 3)(x + 2)}$.

43. $\dfrac{3x^3y + 5x^3 + 3x^2y - 3xy^3 - 14xy^2}{(x - y)(x + y)(x - 2y)(x + 2y)}$.   45. $\dfrac{2 - 3x}{x(x - 1)}$.

47. $\dfrac{3x^4 + 8x^3 + 4x^2 - 5x - 4}{(x + 2)(x + 1)(x - 1)(x^2 + x - 1)}$.   49. $\dfrac{3x^2 - 7x - 32}{x(x - 4)}$.

51. $\dfrac{2x^2 - 17}{x^4 - 16}$.   53. $\dfrac{n + 4}{n - 5}$.   55. $\dfrac{10x^2}{3x - 2}$.   57. $6/(x + 2)$.

59. $3/(x + y)$.   61. $(2 + x)/(2 - x)$.   63. $(a - 3)/(a + 2)$.

65. $x/(x - 1)$.   67. $\dfrac{xy + 2}{xy - y^2}$.   69. $\dfrac{x + y + 1}{2(y - x)}$.

71. $\dfrac{x^2 + 2x - 4}{2x}$.   73. $\dfrac{x + 12}{4x^2 - 5x - 75}$.

Exercise 32

1. 8.   3. 24.   5. 8.   7. 6.   9. 2.   11. $\frac{2}{3}$.
13. -4.   15. 9.   17. 1.   19. $-\frac{1}{5}$.   21. 3.   23. -2.
25. -6.   27. $7, -2$.   29. $2, \frac{3}{5}$.   31. 3.   33. 6.
35. -2.   37. $5, 13$.   39. $\frac{9}{10}$.   41. $-3, -\frac{17}{81}$.

43. $0, 6$.   45. $R = E/I$.   47. $\dfrac{2S + gt^2}{2t}$.   49. Fd^2/m_2.

51. $rl - rs + S$.   53. $\dfrac{RV_1}{V_1 - R}$.   55. $\dfrac{WR - 2PR}{W}$.   57. $\dfrac{S - P}{Pi}$.

59. $\dfrac{en - Cnr}{C}$.

Exercise 33

1. $\frac{2}{5}$ quart.   3. 45 years.   5. 24 feet.   7. 56.   9. -13.
11. 18, 20, 22.   13. 13/23.   15. $-1/3$.   17. $7/12$ or $-20/-15$.
19. 45, 20.   21. $2\frac{7}{19}$ hours.   23. 5 hours, 20 minutes.
25. $6\frac{2}{3}$ days.   27. $60.   29. 3 miles per hour.
31. 60 miles per hour, 40 miles per hour.   33. $10\frac{10}{11}$ miles per hour.
35. 360 miles.   37. 3.7 miles per hour (approximately).

Exercise 34

In Problems 1–7 the independent variable is listed first.
1. t, n.   3. v, t.   5. w, p.   7. $x, 2x^2 + 7x$.   9. b.   11. c.   13. c.   15. b.
In Problems 17–27 the respective values of $f(3), f(1), f(0), f(-1), f(-3)$ are
17. $-21, -7, 0, 7, 21$.   19. $-23, -13, -8, -3, 7$.
21. $-18, -2, 0, -2, -18$.   23. $-6, 0, 0, -2, -12$.
25. $27, 1, 0, -1, -27$.   27. $-42, -4, 0, -6, -48$.
29. $f(a) = a^2 - a^4 = f(-a)$.   31. $f(a) = 8a = -f(-a)$.

33. $f(a) = 2a^3 - 3a = -f(-a)$.   35. $\dfrac{a}{1 - a^2}$.

37. (a) d, the diameter; (b) c, the circumference; (c) $f(5) = 5\pi$, the circumference of a circle whose diameter is 5.   39. $A = 110 - 5n$.   41. (a) $p = 110 + n/2$; (b) $f(22) = 121$, the blood pressure of a person of age 22.

Exercise 35

Problem	x	-4	-3	-2	-1	0	1	2	3	4	If x is increased by 1, then
1.	y	-12	-9	-6	-3	0	3	6	9	12	y is increased by 3
3.	y	-14	-10	-6	-2	2	6	10	14	18	y is increased by 4
5.	y	19	16	13	10	7	4	1	-2	-5	y is decreased by 3

7. -0.017.   9. 765.   11. $-$85.   13. Decreasing, negative.
15. Decreasing, negative.   17. Increasing, positive.
19. Decreasing, negative.

21.

x	1	3	5	7	9	11
y	6	54	150	294	486	726

23. 48.   25.

x	1	3	5	7	9
y	1	27	125	343	729

27. 49.   29. 450 gallons.

31. $\Delta g = -30$ for each interval. It is constant.   33. Negative. This means that as time increases, the amount of water in the tank decreases.
35. $y = 4x$.   37. $y = 2x + 10$.   39. $y = -x$.   41. 4.   43. 41.   45. 5.14.

Exercise 36

3. $E(3, 1)$, $F(1, 3)$, $G(2, -2)$, $H(0, -3)$, $I(-2, 2)$, $J(-2, 0)$, $K(-3, -2)$, $L(4, -1)$.   5. (a) rectangle; (b) isosceles triangle; (c) trapezoid; (d) parallelogram; (e) square.   7. (c) Yes.   9. (c) The ordinate of each point is 1 greater than its corresponding abscissa; (d) The point will be on the line; (e) $y = x + 1$.   11. (a) 13; (b) The lengths of the three sides are 10, $\sqrt{101}$, and $\sqrt{97}$; (c) The lengths of the three sides are $\sqrt{65}$, $\sqrt{65}$, and $\sqrt{52}$; (d) $\sqrt{(y_2 - y_1)^2 + (x_2 - x_1)^2}$.   13. (a) One-half their sum; (b) one-half their sum; (c) $\left(\dfrac{x_1 + x_2}{2}, \dfrac{y_1 + y_2}{2}\right)$.

Exercise 37

1. $m = 2, b = 3$.   3. $m = \frac{1}{2}, b = 0$.   5. $m = \frac{2}{3}, b = 2$.
7. $m = 1, b = 0$.   9. $m = 0, b = 6$.   11. (a) 6; (b) $\Delta y/\Delta x = 6$;
(c) $\Delta y/\Delta x$ is equal to the slope.   13. (a) 4; (b) 10; (c) 6, yes.
15. (5, 0), (0, 10).   17. (2, 0), (0, -4).   19. (12, 0), (0, -3).
21. (20/7, 0), (0, -10).   23. Parallel.   25. Parallel.
27. Parallel.   29. Coincide.   31. Coincide.

33. (a)

x	1	2	3	4	5	6	7
y	-8	-3	0	1	0	-3	-8

; (c) $x = 3$, $x = 5$; (d) equal.

35.

x	2	4	6	8	10	12
y	32	12	0	-4	0	12

; vertex is at (8, -4).

37.

x	-1	0	1	2	3	4	5	6	7	8
y	0	-7	-12	-15	-16	-15	-12	-7	0	9

; vertex is at (3, -16).

39.

x	-3	-2	-1	0	1	2	3	4	5
y	-7	0	5	8	9	8	5	0	-7

; vertex is at (1, 9).

41. (4, -12).   43. (2, -8).   45. (0,0).   47. (-3, 27).

Exercise 38

1. (4, 1).   3. (8, -1).   5. Infinite number of solutions.
7. No solution.   9. (-7, 4).   11. No solution.   13. (4,7).

15. No solution.   17. (3, −3).   19. (−3, 6), (2, −4).
21. (3, 3).   23. (3.4, 0), (−1.4, 0).   25. No solution.
27. (4.4, 3.4), (−2.4, −3,4).   29. 0, 2, −3.   31. (b) $x = 1 \pm 3 \sqrt{-1}$.

Exercise 39

1. (a) 2; (b) 11.   3. −8, minimum.   5. 25, maximum.
7. −12.5, minimum.   9. −4.25, minimum.   11. $x = 0$, maximum;
$x = 2$, minimum.   13. $x = 1$, minimum; $x = -\frac{1}{3}$, maximum.
15. −1.   17. Minimum. Parabola is concave up.   19. Minimum.
Parabola is concave up.   21. Minimum. Parabola is concave up.
23. (a) $12 - x$; (b) $y = x(12 - x)$; (c) 6.   25. 3 inches deep, 6 inches
wide.   27. 36 feet.   29. 13 cents.

Exercise 40

1. $y < (3x + 4)/2$.   3. $y < (2x - 6)/3$.   5. $y > 4 - x$.   7. $y < 2x + 1$.
9. $y > 6 - 3x$.   11. $y < x$.   13. $x < 8$.   15. $x > 3$.
17. $x > \frac{13}{2}$.   19. To the right of the straight line graph $x = y - 2$.

Exercise 41

1. Exact.   3. Approximate.   5. Exact.   7. Approximate.
9. Approximate.   11. Approximate.   13. Approximate.
15. 10 is exact, "can" is approximate.   17. 1 is exact, "lump" is approximate.
19. 4 is exact, "quart bottle" is approximate.   21. Exact.   23. 125 is exact,
"kilowatt-hour" is approximate.   25. Approximate.   27. Indirect.   29. Direct.
31. Direct.   33. Indirect.   35. Direct.

Exercise 42

1. 0.005 pound; 0.0020; 2.535 pounds and 2.545 pounds.   3. 0.05 inch; 0.0057;
8.65 inches and 8.75 inches.   5. 50 miles per second; 0.00027; 186,150 miles
per second and 186,250 miles per second.   7. 5 feet; 0.00095; 5275 feet and
5285 feet.   9. 0.05 bushel; 0.0040; 12.35 bushels and 12.45 bushels.   11. 0.00005
inch; 0.00013; 0.39365 inch and 0.39375 inch.   13. 0.00005 inch; 0.000016;
3.14155 inches and 3.14165 inches.   15. 0.000 005 inch; 0.020; 0.000245 inch and
0.000255 inch.   17. 2, 4.   19. 3.   21. 1, 8, 4, 3.   23. 2, 4, 8, 2.
25. 1, 0, 4.   27. 2, 0, 0, 0.   29. 6, 0, 0, 0.   31. Not the same.
33. Same.   35. Not the same.   37. Same.   39. Same.
41. Not the same.   43. Same.   45. Same.   47. Not the same.
49. Same.

Exercise 43

1. 2.41.   3. 62.3.   5. 82.5.   7. 47.4.   9. 8.78
11. 1080.   13. 1.45.   15. 1.01.   17. 4.99.   19. 1.38.
21. 3918.8.   23. 8.72.   25. 0.015.   27. 4.69.

29. 2440.   31. 0.011.   33. 36,900.   35. 104.63.
37. 0.0173.   39. 1000.   41. 4.6.   43. 1.7 inches.
45. 15.8 square inches.   47. 4.500 inches.   49. 215 gallons.

Exercise 44

1. 1.   3. 1.   5. $3a + 1$.   7. 0.   9. $1 - 3a^2$.   11. 2.
13. 3.   15. -2.   17. 7.   19. -5.   21. $3a$.   23. 3.
25. 4.   27. 8.   29. 9.   31. 27.   33. 27.   35. 4.
37. $27x^3$.   39. $4a^2$.   41. $\frac{1}{9}$.   43. 8.   45. $\frac{27}{8}$.   47. $\frac{1}{2}$.
49. $\frac{1}{9}$.   51. $\frac{1}{9}$.   53. 4.   55. 2.   57. $\frac{4}{3}$.
59. 2.   61. $3/a$.   63. a^2c/b.   65. $3a^2/b$.   67. y^2/x.
69. $2a/x^2$.   71. $a^2/2$.   73. ac/b^2.   75. $2b/a$.   77. $(x/2)^2$.
79. $(x^3/8)^{1/3}$.   83. $11/10^5$.   85. $a + b$.   87. $\dfrac{a^2 + b^2}{a^2b^2}$.   89. $\frac{3}{32}$.

Exercise 45

1. 3.57×10^5.   3. 1.47×10^7.   5. 2.3×10^{14}.   7. 1.1×10^{-5}.
9. 3×10^{-8}.   11. 1.60×10^{-12}.   13. 0.000 030 30.
15. 0.000 021 42.   17. 0.000 000 000 000 027 78.
19. 10,000,000,000,000.   21. 0.000 000 073 76.   23. 1,013,250.
25. 20.   27. 1100.   29. 0.0004.   31. 8.   33. 0.4.

Exercise 46

1. $\text{Log}_9\, 81 = 2$.   3. $\text{Log}_2\, 16 = 4$.   5. $\text{Log}_{64}\, 8 = \frac{1}{2}$.   7. $\text{Log}_6\, 1 = 0$.
9. $\text{Log}_5\, 125 = 3$.   11. $\text{Log}_2\, \frac{1}{8} = -3$.   13. $5^2 = 25$.   15. $16^{1/2} = 4$.
17. $5^{-2} = \frac{1}{25}$.   19. $9^0 = 1$.   21. $4^{5/2} = 32$.   23. $8^1 = 8$.
25. 3.   27. 4.   29. -1.   31. $\frac{4}{3}$.   33. $\frac{1}{2}$.   35. $\frac{1}{64}$.
37. 10.   39. 16.   41. 0.001.   43. (a) 3; (b) 2; (c) 5.
45. (a) -1; (b) 3; (c) 2.   47. (a) 5; (b) 1; (c) 4.
49. (a) 3; (b) 2; (c) 1.   51. (a) 0; (b) 0; (c) 0.   53. Yes. The logarithm of a positive number less than 1 to a positive base is negative, for example, $\text{log}_5\, \frac{1}{25} = -2$.

Exercise 47

1. $a + b$.   3. $b - a$.   5. $\frac{1}{2}b$.   7. $\frac{1}{3}(a - b)$.   9. $\frac{1}{2}a - 2b$.
11. $-a$.   13. 0.6020.   15. 1.4771.   17. 5.4467.   19. 0.9209.
21. 0.9376.   23. 0.2000.   25. 0.3153.   27. 0.5340.

29. 1.1520.   31. $\text{Log}_3\, (x^4)$.   3. $\text{Log}_{10}\, (U^3 V^2)$.   35. $\text{Log}_5\left(\dfrac{x^{1/3}}{y^2}\right)$.

37. $\text{Log}\, x = \log a + \log b$.   39. $\text{Log}\, x = 3 \log a + 2 \log b$.
41. $\text{Log}\, x = \log b - \log a$.   43. $\text{Log}\, x = \frac{1}{2} \log a - \log b$.
45. $\text{Log}\, x = \frac{1}{3} \log a - \frac{1}{2} \log b$.   47. $\text{Log}\, x = 2 \log a + \frac{1}{2} \log b$.
51. $\text{Log}\, y = \log p + \log r + \log t$.   53. $\text{Log}\, y = \log p + n \log (1.06)$.

Exercise 48

1. 3.   3. 1.   5. 4.   7. 1.   9. 0.   11. −4.   13. −2.
15. −3.   17. 409,000.   19. 0.0409.   21. 0.000409.
23. 0.9143 − 2.   25. 1.0253.   27. 0.8451 − 3.   29. 0.4871.
31. 0.9685 − 2.   33. 4.3032.   35. 1.9894.   37. 4.0000.
39. 0.6096 − 6.   41. 4.12.   43. 7550.   45. 0.135.
47. 100.   49. 0.545.   51. 0.000 002 51.   53. 0.000759.
55. 30,900.   57. 64,000.   59. 27.8.   61. 6.32.
63. 23,900,000.   65. 73,000,000.

Exercise 49

1. 333.   3. 9.14.   5. 0.000 043 7.   7. 0.000206.   9. 0.342.
11. 5540.   13. 0.0675.   15. 0.592.   17. 0.0389.   19. 1.51.
21. 9.40.   23. 0.781.   25. 2.27×10^{-17}.   27. 0.000978.
29. 1520.   31. 77.6.   33. 478.   35. 361.   37. 217.
39. 600.   41. $134.   43. $179.   45. $432.
47. 4.89 seconds.   49. 11.7 square inches.

Exercise 50

1. 3.6108.   3. 4.7651.   5. 2.4470.   7. 0.9174.   9. 1.9551.
11. 0.003753.   13. 734,700.   15. 0.4349.   17. 30,540.
19. 0.000 053 33.   21. 6.528.   23. 0.08862.   25. 0.2795.
27. 0.5912.   29. 1.390.   31. 7.668.   33. 1.531×10^{11}.
35. 351.4 square feet.   37. 0.8251 ohm.

Exercise 51

1. Does not.   3. Does not.   5. Does not.   7. Does.
9. Does.   11. Does not.   13. Does.   15. 8/3.   17. 2/7.
19. 9/4.   21. 1/9.   23. 24.   25. 40/3.   27. 1/4.
29. 5/12.   31. 1/14.   33. 10/3.   35. 15/17.   37. 24/5.
39. 6, −6.   41. 27/7.   43. 16, −16.   45. 4, −4.
47. $3 \sqrt{15}, -3 \sqrt{15}$.   49. $\sqrt{ab}, -\sqrt{ab}$.   51. $ab, -ab$.   53. $4a^2, -4a^2$.
55. Multiply each member by b/c.   57. Add 1 to each member, and then
rewrite each member as a single fraction.   59. 2/8 = 3/12 (alternation);
3/2 = 12/8 (inversion); 5/3 = 20/12 (composition); −1/3 = −4/12 (division).   61. 562.5 gallons.   63. 0.042 pound.   65. 10 pounds of tin,
15 pounds of lead.

Exercise 52

1. $C = kr$; $C_1/C_2 = r_1/r_2$; $k = \pi$.   3. $w = kl$; $w_1/w_2 = l_1/l_2$;
k represents the weight per unit length.   5. $h = kt$; $h_1/h_2 = t_1/t_2$.
7. $A = ks^2$; $A_1/A_2 = (s_1/s_2)^2$; $k = \sqrt{3}/4$.   9. $A = kd^2$; $A_1/A_2 = (d_1/d_2)^2$.
11. $V = kr^3$; $V_1/V_2 = (r_1/r_2)^3$; $k = \frac{4}{3}\pi$.   13. $P = kv^3$; $P_1/P_2 = (V_1/V_2)^3$.

15. $L = kd^4; L_1/L_2 = (d_1/d_2)^4.$ 17. $t^2 = kd^3; (t_1/t_2)^2 = (d_1/d_2)^3.$

19. $k = 4; y = 4x; y = 16.$ 21. $k = 1; y = x^2; y = 16.$

23. $k = \sqrt{2}; y = \sqrt{2x}; x = 16.$ 25. $k = \frac{15}{8}; y = \dfrac{15}{8} x^{\frac{3}{2}}; y = 15.$

27. (a) $\frac{1}{6}$; (b) 2.   29. (a) 63; (b) 4.   31. (a) y is 4 times as large;
(b) y is 9 times as large; (c) y is $\frac{1}{4}$ as large; (d) y is $\frac{1}{16}$ as large.   33. No; no.
35. (a) $R = 0.00253\ l$, (b) 0.0506 ohm; (c) 395 feet.   37. (a) $P = \frac{1}{450}v^2$;
(b) $4\frac{1}{2}$ pounds per square foot; (c) 60 miles per hour.   39. About 210 feet.
41. About 23,000 horsepower.

Exercise 53

1. $y_1 = k/x_1$, $y_2 = k/x_2$; therefore $y_1/y_2 = x_2/x_1.$
3. $w = k/s^2; w_1/w_2 = (s_2/s_1)^2.$ 5. $I = k/d^2; I_1/I_2 = (d_2/d_1)^2.$
7. $g = k/s^2; g_1/g_2 = (s_2/s_1)^2.$ 9. $T = k/V; T_1/T_2 = V_2/V_1.$
11. $h = k/r^2; h_1/h_2 = (r_2/r_1)^2; k = 1/\pi.$ 13. $A = k/h; A_1/A_2 = h_2/h_1.$
15. $k = 16; y = 16/x; y = 4.$ 17. $k = 54; y = 54/x^2; y = \frac{27}{2}.$
19. $k = 15; y = 15/\sqrt{x}; y = \frac{5}{2}.$ 21. (a) 30; (b) 5.
23. (a) $62\frac{1}{2}$; (b) 625.   25. No; no.
27. (a) y is $\sqrt{2}/2$ times as large; (b) y is $\sqrt{3}/3$ times as large; (c) y is $\sqrt{2}$
times as large; (d) y is twice as large.   29. About 130 pounds.   31. About
14 feet.   33. 3.52 inches.

Exercise 54

1. $p = kdt; \dfrac{p_1}{p_2} = \dfrac{d_1 t_1}{d_2 t_2}.$ 3. $R = \dfrac{kl}{d^2}; \dfrac{R_1}{R_2} = \dfrac{l_1}{l_2}\left(\dfrac{d_2}{d_1}\right)^2.$

5. $L = kAv^2; \dfrac{L_1}{L_2} = \dfrac{A_1}{A_2}\left(\dfrac{v_1}{v_2}\right)^2.$ 7. 1.   9. Price per gallon.

11. $E = \dfrac{kI}{d^2}; \dfrac{E_1}{E_2} = \dfrac{I_1}{I_2}\left(\dfrac{d_2}{d_1}\right)^2.$ 13. $k = 4; z = 4xy; z = 60.$

15. $k = 7; z = 7x/y; z = 3.$ 17. $k = 18; z = 18x/\sqrt{y}; z = 90.$
19. $k = 10; z = 10xy/w^2; z = 27.$ 21. (a) 15; (b) 4.
23. (a) 36; (b) 6.   25. 19.6 cubic feet.
27. 105 pounds.   29. 8.48 inches.

Exercise 55

1. $\{1, 2, 3, 4, 5, 6, 7, 8, 9, 10, 11, 12\}.$ 3. $\{8, 9, 10, 11, 12\}.$ 5. $\{3, 8\}.$
7. P is the set of squares of all positive integers.
9. Q is the set of all negative odd integers greater than $-10.$
11. G is the set of all triangles.   13. True.   15. False.
17. True.   19. True.   21. True.   23. 3; $\{p, q\}, \{p, r\}, \{q, r\}.$
25. $\phi, \{a\}, \{b\}, \{c\}, \{a, b\}, \{a, c\}, \{b, c\}, \{a, b, c\}.$ 27. 8.   29. $2^n.$
31. $\{-3, -1, 0\}.$ 33. $\{-1, 0, 1, 3\}.$ 35. $\{-3, -1, 0, 1, 3\}.$ 37. $\phi.$
51. $\{-1\}.$ 53. $\{3\}.$ 55. $\{\frac{1}{2}, \frac{2}{3}\}.$ 57. $\{-1, -2, -3, \ldots\}.$

Exercise 56

1. $\{(a, r)\}$.   3. $\{(p, r), (p, s), (q, r), (q, s)\}$.   5. $\{(0, 1), (0, 3),$
$(0, 5), (2, 1), (2, 3), (2, 5), (4, 1), (4, 3), (4, 5)\}$.   7. mn.
9. $\{(u, u), (u, v), (v, v), (v, u)\}$.   11. $\{(p, p), (p, q), (p, r), (p, s), (q, p),$
$(q, q), (q, r), (q, s), (r, p), (r, q), (r, r), (r, s), (s, p), (s, q), (s, r), (s, s)\}$.
13. $\{(1, 1), (1, 2), (1, 3), (1, 4), (1, 5), (1, 6), (2, 1), (2, 2), (2, 3), (2, 4), \ldots\}$.
15. 400.   17. 20.   19. $\{(1, 1), (3, 3), (5, 5), (7, 7)\}$.
21. $\{(1, 3), (1, 5), (1, 7), (3, 5), (3, 7), (5, 7)\}$.   23. $\{(1, 3), (3, 5), (5, 7)\}$.
25. $\{(-3, -3), (-3, -2), (-3, -1), (-3, 0), (-3, 1), (-3, 2), (-3, 3),$
$(-2, -2), (-2, -1), (-2, 0), (-2, 1), (-2, 2), (-2, 3), (-1, -1), (-1, 0),$
$(-1, 1), (-1, 2), (-1, 3), (0, 0), (0, 1), (0, 2), (0, 3), (1, 1), (1, 2), (1, 3), (2, 2),$
$(2, 3), (3, 3)\}$.
27. $\{(-3, -3), (-3, -2), (-3, -1), (-3, 0), (-3, 1), (-3, 2), (-3, 3),$
$(-2, -2), (-2, -1), (-2, 0), (-2, 1), (-2, 2), (-2, 3), (-1, 0), (-1, 1),$
$(-1, 2), (-1, 3), (0, 2), (0, 3)\}$.
29. Domain = range = $\{0, 1, 2, 3, 4, 5\}$.
31. Domain = range = $\{0, 1, 2, 3, 4, 5\}$.
39. Function.   41. Is not a function.   43. Is not a function.
45. Is not a function.

Exercise 57

1. $\{-3, -2, -1, 0\}$.   3. $\{1, 3\}$.   5. $\{2, 4\}$.   7. $\{b, c, d\}$.
9. ϕ.   11. $\{a, e, f, g, h\}$.   13. Their union is the set of real numbers; their
intersection is the empty set ϕ.   21. $\{x \in U \mid x \le 3\}$.   23. $\{x \in U \mid x \ne 5\}$.
47. Universal set U.   49. The point of intersection of the bisectors of angle B
and angle C.

Exercise 58

1. $G \cup K = \{-2, -1, 0, 1, 2, 3, 4, 5\} = K \cup G; G \cap K = \{1, 2, 3\} = K \cap G;$
$\overline{G \cup K} = \{-5, -4, -3\} = \overline{G} \cap \overline{K}; \overline{G \cap K} = \{-5, -4, -3, -2, -1, 0\} =$
$\overline{G} \cup \overline{K}$.
3. $R \cup S = \{a, b, c, d, e\} = S \cup R; R \cap S = \{a, c, e\} = S \cap R; \overline{R \cup S} =$
$\{f\} = \overline{R} \cap \overline{S}; \overline{R \cap S} = \{b, d, f\} = \overline{R} \cup \overline{S}$.
9. The union of any set with the empty set is the set itself.   11. The intersection
of any set with the universal set is the universal set.   13. (a) B; (b) A.
15. (a) A or B; (b) A or B.   17. $(A \cap B) \cap C = A \cap (B \cap C)$.
19. $A \cap U = A$.   21. $\overline{A \cap B} = \overline{A} \cup \overline{B}$.
23. ϕ.   25. $\overline{A}$.   27. ϕ.   29. Q.   31. $G \cup H$.   33. B.
35. R.   37. ϕ.   39. B.   41. $\overline{B} \cup C$.   43. P.   45. $P \cap \overline{Q}$.

Index